LAKE VIEWS

LAKE VIEWS

This World and the Universe

STEVEN WEINBERG

The Belknap Press of
Harvard University Press
Cambridge, Massachusetts
London, England
2009

Library of Congress Cataloging-in-Publication Data

Weinberg, Steven, 1933–
Lake views : this world and the universe / Steven Weinberg.
p. cm.
ISBN 978-0-674-03515-7 (alk. paper)
1. Science. I. Title.
Q171.W4194 2009
500—dc22   200917607

To Louise, Elizabeth, and Gabrielle

Contents

Preface

This book is a collection of essays published in various periodicals and books in the years 2000 to 2008. Its title reflects the fact that these essays, like my research articles, were all written in my study at home, on the shore of Lake Austin, at a desk by a window overlooking the lake.

It isn't really a lake. Like virtually all so-called lakes in Texas, Lake Austin is a segment of a dammed up river, in this case the Texas Colorado. It is twenty miles long and, at my house, less than a quarter of a mile wide—so narrow, that the Lower Colorado River Authority requires boats on the lake to keep to the right. Especially in summer, one can see party boats and ski boats going up or down the lake, and hear their engines and their music. So, though most of my work as a physicist is mathematical and impersonal, through my window I am continually reminded, like some lakeside Lady of Shalott, of the human world outside.

The twenty-five essays in this collection are presented here in chronological order. They express my views on topics ranging from problems of cosmology to issues of this world—military, political, and religious. Although many of these topics range far outside the bounds of science, in one way or another they all reflect my experience as a theoretical physicist. Like the essays in my earlier Harvard Press collection, *Facing Up*, they express a viewpoint that is rationalist, reductionist, realist, and devoutly secular.

Nine of the essays presented here appeared originally in the *New York Review of Books*. Of the other essays, most were published in periodicals or books intended for the general reader. But a book review and two other articles appeared in a journal, *Physics Today*, published monthly for physicists by the American

Institute of Physics. *Physics Today* aims to keep the whole physics community up to date on anything related to physics, so its articles avoid the technical language of specialized research articles. Even so, on rereading these articles I saw that in them I had sometimes casually used words that would be generally familiar to physicists, but not to other readers. Where this was the case, I have added explanations of these terms, or found other language.

One other article, "Living in the Multiverse," was written for theoretical physicists, and I have deleted the mathematical parts. The remaining essays in this collection are presented here pretty much as they originally appeared, with just a little rewriting where clarification or correction seemed necessary or where I wanted to avoid repeating myself. Where I have had afterthoughts about the topics discussed, I have generally included them as footnotes, labeled "Added note." As in *Facing Up,* I have added a new introduction to each essay, explaining how it came to be written. For some essays I have also added an afterword to bring the subject of the essay up to date.

In a few of these articles I need to refer to numbers that are very large or very small, and for that purpose I use a notation that will probably be familiar to most but perhaps not to all readers, writing these numbers as powers of 10 or $\frac{1}{10}$. For instance, 10^{56} is the product of fifty-six tens, or a one followed by fifty-six zeroes, and 10^{-33} is the product of thirty-three factors of $\frac{1}{10}$, or a decimal point followed by thirty-two zeroes and a one.

I am grateful again to Michael Fisher of Harvard University Press for suggesting the publication of the original collection of my articles, *Facing Up,* and for his help now in publishing this second collection. Thanks are due to Terry Riley and Jan Duffy for finding many books and articles and old files. I gratefully acknowledge the help of numerous colleagues at the University of Texas who provided information about special topics. For suggestions that I think have greatly improved these articles I owe thanks to many friends and editors, especially Robert Silvers of the *New York Review of Books.*

LAKE VIEWS

1

Waiting for a Final Theory

The end of the twentieth century prompted magazines and newspapers to indulge in a good deal of guesswork about the future. As part of this prophetic effort, *Time* magazine asked me to assess how far we yet have to go in understanding the fundamental laws of nature. My answer: pretty far. This brief essay was published by *Time* in April 2000.

The twentieth century was quite a time for physicists. By the mid-1970s we had in hand the so-called Standard Model, a theory that accurately describes the forces and particles we observe in our laboratories and that provides a basis for understanding virtually everything else in physical science.

No, we don't actually understand everything—there are many things, from the turbulence of ocean currents to the folding of protein molecules, which cannot be understood without new insights and radical improvements in our methods of calculation. They will provide plenty of interesting continued employment for theorists and experimenters for the foreseeable future. But no new free-standing scientific principles will be needed to understand these phenomena. The Standard Model provides all the fundamental principles we need for this.[1]

There is one force, though, that is not covered by the Standard Model: the force of gravity. Einstein's General Theory of Relativ-

---

1. This is a very brief statement of a reductionist viewpoint described in more detail in my book *Dreams of a Final Theory* (Pantheon, 1992).

ity gives a good account of gravitation at ordinary distances, and if we like, we can tack it onto the Standard Model. But serious mathematical inconsistencies turn up when we try to apply it to particles separated by tiny distances—distances about 10 million billion times smaller than those probed in the most powerful particle accelerators.

Even apart from its problems in describing gravitation, however, the Standard Model in its present form has too many arbitrary features. Its equations contain too many constants of nature—such as the masses of the elementary particles and the strength of the fundamental unit of electric charge—that are given specific numerical values for no other reason than that these values seem to work. In writing these equations, physicists simply plugged in whatever values made the predictions of the theory agree with experimental results.

There are reasons to believe that these two problems are really the same problem. That is, we think that when we learn how to make a mathematically consistent theory that governs both gravitation and the forces already described by the Standard Model, all those seemingly arbitrary properties will turn out to be what they are because this is the only way that the theory can be mathematically consistent.

One clue that this should be true is a calculation showing that, although the strengths of the various forces seem very different when measured in our laboratories, they would all be equal if they could be measured at tiny distances—distances close to those at which the above-mentioned inconsistencies begin to show up.

Theorists have even identified a candidate for a consistent unified theory of gravitation and all the other forces: superstring theory. In some versions, it proposes that what appear to us as particles are stringy loops or lines that exist in a space-time with ten dimensions. But we don't yet understand all the principles of this theory, and even if we did, we would not know how to use the theory to make predictions that we can test in the laboratory.

Such an understanding could be achieved tomorrow by some bright graduate student, or it might just as well take another century or so. It may be accomplished by pure mathematical deduction from some fundamental new physical principle that just happens to occur to someone, but it is more likely to need the inspiration of new experimental discoveries.

We would like to be able to judge the correctness of a new fundamental theory by making measurements of what happens at scales ten million billion times smaller than those probed in today's laboratories, but this may always be impossible. With any technology we can now imagine, measurements like those would take more than the economic resources of the whole human race.

Even without new experiments, it may be possible to judge a final theory by whether it explains all the apparently arbitrary aspects of the Standard Model. But there are explanations and explanations. We would not be satisfied with a theory that explains the Standard Model in terms of something complicated and arbitrary, in the way astronomers before Kepler explained the motions of planets by piling epicycles upon epicycles.

To qualify as an explanation, a fundamental theory has to be simple—not necessarily a few short equations, but equations that are based on a simple physical principle, in the way that the equations of General Relativity are based on the principle that gravitation is an effect of the curvature of space-time. And the theory also has to be compelling—it has to give us the feeling that it could scarcely be different from what it is.

When at last we have a simple, compelling, mathematically consistent theory of gravitation and other forces that explains all the apparently arbitrary features of the Standard Model, it will be a good bet that this theory really is final. Our description of nature has become increasingly simple. More and more is being explained by fewer and fewer fundamental principles. But simplicity can't increase without limit. It seems likely that the next major theory that we settle on will be so simple that no further simplification would be possible.

The discovery of a final theory is not going to help us cure cancer or understand consciousness, however. We probably already know all the fundamental physics we need for these tasks. The branch of science in which a final theory is likely to have its greatest impact is cosmology. We have pretty good confidence in the ability of the Standard Model to trace the present expansion of the universe back to about a billionth of a second after its supposed start. But when we try to understand what happened earlier than that, we run into the limitations of the model, especially its silence on the behavior of gravitation at very short distances.

The final theory will let us answer the deepest questions of cosmology: Was there a beginning to the present expansion of the universe? What determined the conditions at the beginning? And is what we call our universe, the expanding cloud of matter and radiation extending billions of light-years in all directions, really all there is, or is it only one part of a much larger "multiverse," in which the expansion we see is just a local episode?

The discovery of a final theory could have a cultural influence as well, one comparable to what was felt at the birth of modern science. It has been said that the spread of the scientific spirit in the seventeenth and eighteenth centuries was one of the things that stopped the burning of witches. Learning how the universe is governed by the impersonal principles of a final theory may not end mankind's persistent superstitions, but at least it will leave them a little less room.

Since 2000 neither superstring theory nor the multiverse idea has yet achieved the kind of success that would establish their permanent place in science, but they have in a sense come together. In 2001 it was found (by Steven Giddings, Shamit Kachru, and Joseph Polchinski at the University of California at Santa Barbara) that the equations of superstring theory have an enormous number of solutions. Leonard Susskind of Stanford coined the term "string landscape" for this multiplicity of solutions, in analogy with the landscape of possible orientations of protein molecules found in theoretical biochemistry. In accordance with earlier ideas of Andrei

Linde at Stanford about chaotic inflation, each of these solutions potentially describes a different sort of big bang cosmology. (This is the subject of article 19 in this collection.) The exploration of the string landscape and its cosmological implications is a formidable task, and has barely begun. Indeed, the distance we still have to go in understanding the fundamental laws of nature seems even greater in 2009 than it did in 2000.

2

Can Science Explain Everything? Anything?

———————

This essay, based on a talk given at Amherst College in October 2000, is one of my occasional reluctant ventures into the philosophy of science. I generally feel diffident in writing about philosophy, as there are smart people who do this full time. Still, it seems to me that the task of the philosopher of science is not to tell scientists how to work, but to describe what scientists are doing at their work, so it might be useful for philosophers to hear from time to time what a working scientist thinks that he or she is doing. But I had another reason for accepting when the philosophers at Amherst College invited me to give a talk on the meaning of explanation in science. I live in Texas, and like it here, but I do sometimes miss the autumn in New England, where I used to live. This trip was not a disappointment. On the way from Boston to Amherst, I drove past countless maples and oaks whose red and yellow leaves were lit up by the October sunshine. I think I enjoyed the drive more than the Amherst audience could have enjoyed my talk.

This essay was published in the *New York Review of Books* in May 2001, and republished in *The Best American Science Writing 2002*.

———————

One evening a few years ago I was with some other faculty members at the University of Texas, telling a group of undergraduates about work in our respective disciplines. I outlined the great progress we physicists had made in explaining what was known experimentally about elementary particles and fields—how when I was a student I had to learn a large variety of miscellaneous facts about particles, forces, and symmetries; how in the decade from the mid-1960s to the mid-1970s all these odds and ends were explained in what is now called the Standard Model of elementary particles; how we learned that these miscellaneous facts about particles and forces could be deduced mathematically from a few fairly simple principles; and how a great collective *Aha!* then went out from the community of physicists.

After my remarks, a faculty colleague (a scientist, but not a particle physicist) commented, "Well, of course, you know science does not really explain things—it just describes them." I had heard this remark before, but now it took me aback, because I had thought that we had been doing a pretty good job of explaining the observed properties of elementary particles and forces, not just describing them.

I think that my colleague's remark may have come from a kind of positivistic angst that was widespread among philosophers of science in the period between the world wars. Ludwig Wittgenstein famously remarked that "at the basis of the whole modern view of the world lies the illusion that the so-called laws of nature are the explanations of natural phenomena."

It might be supposed that something is explained when we find its cause, but an influential 1913 paper by Bertrand Russell[1] had argued that "the word 'cause' is so inextricably bound up with misleading associations as to make its complete extrusion from the philosophical vocabulary desirable." This left philosophers like Wittgenstein with only one candidate for a distinction between explanation and description, one that is teleological, defining an explanation as a statement of the purpose of the thing explained.

E. M. Forster's novel *Where Angels Fear to Tread* gives a good example of teleology making the difference between description

---

1. "On the Notion of Cause," reprinted in *Mysticism and Logic* (Doubleday, 1957), p. 174.

and explanation. Philip is trying to find out why his friend Caroline helped to bring about a marriage between Philip's sister and a young Italian man of whom Philip's family disapproves. After Caroline reports all the conversations she had with Philip's sister, Philip says, "What you have given me is a description, not an explanation." Everyone knows what Philip means by this—in asking for an explanation, he wants to learn Caroline's purposes. There is no purpose revealed in the laws of nature, and not knowing any other way of distinguishing description and explanation, Wittgenstein and my friend had concluded that these laws could not be explanations. Perhaps some of those who say that science describes but does not explain mean also to compare science unfavorably with theology, which they imagine to explain things by reference to some sort of divine purpose, a task declined by science.

This mode of reasoning seems to me wrong not only substantively, but also procedurally. It is not the job of philosophers or anyone else to dictate meanings of words different from the meanings in general use. Rather than argue that scientists are incorrect when they say, as they commonly do, that in their work they are explaining things, philosophers who care about the meaning of explanation in science should try to understand what it is that scientists are doing when they say they are explaining something. If I had to give an a priori definition of explanation in physics I would say, "Explanation in physics is what physicists have been doing when they say *Aha!*" But a priori definitions (including this one) are not much use.

As far as I can tell, this has become well understood by philosophers of science at least since World War II. There is a large modern literature on the nature of explanation, by philosophers like Peter Achinstein, Carl Hempel, Philip Kitcher, and Wesley Salmon. From what I have read in this literature, I gather that philosophers are now going about this the right way: they are trying to develop an answer to the question "What is it that scientists do when they explain something?" by looking at what scientists are actually doing when they *say* they are explaining something.

Scientists who do pure rather than applied research commonly tell the public and funding agencies that their mission is the explanation of some thing or other; so the task of clarifying the nature of explanation can be pretty important to them, as well as to philosophers. This task seems to me to be a bit easier in physics (and chemistry) than in other sciences, because philosophers of science have had trouble with the question of what is meant by an explanation of an *event* (note Wittgenstein's reference to "natural phenomena") while physicists are interested in the explanation of regularities, of physical principles, rather than of individual events.

Biologists, meteorologists, historians, and so on are concerned with the causes of individual events, such as the extinction of the dinosaurs, the blizzard of 1888, the French Revolution, etc., while a physicist only becomes interested in an event, like the fogging of Becquerel's photographic plates that in 1897 were left in the vicinity of a salt of uranium, when the event reveals a regularity of nature, such as the instability of the uranium atom. Philip Kitcher has tried to revive the idea that the way to explain an event is by reference to its cause, but which of the infinite number of things that could affect an event should be regarded as its cause?[2]

Within the limited context of physics, I think one can give an answer of sorts to the problem of distinguishing explanation from mere description, which captures what physicists mean when they say that they have explained some regularity. The answer is that

2. There is an example of the difficulty of explaining events in terms of causes that is much cited by philosophers. Suppose it is discovered that the mayor has paresis. Is this explained by the fact that the mayor had an untreated case of syphilis some years earlier? The trouble with this explanation is that most people with untreated syphilis do not in fact get paresis. If you could trace the sequence of events that led from the syphilis to the paresis, you would discover a great many other things that played an essential role—perhaps a spirochete wiggled one way rather than another way, perhaps the mayor also had some vitamin deficiency—who knows? And yet we feel that in a sense the mayor's syphilis *is* the explanation of his paresis. Perhaps this is because the syphilis is the most dramatic of the many causes that led to the effect, and it certainly is the one that would be most relevant politically.

we explain a physical principle when we show that it can be deduced from a more fundamental physical principle. Unfortunately, to paraphrase something that Mary McCarthy once said about a book by Lillian Hellman, every word in this definition has a questionable meaning, including "we" and "a." But here I will focus on the three words that I think present the greatest difficulties: the words "fundamental," "deduced," and "principle."

The troublesome word "fundamental" can't be left out of this definition, because deduction itself doesn't carry a sense of direction; it often works both ways. The best example I know is provided by the relation between the laws of Newton and the laws of Kepler. Everyone knows that Newton discovered not only a law that says the force of gravity decreases with the inverse square of the distance, but also a law of motion that tells how bodies move under the influence of any sort of force. Somewhat earlier, Kepler had described three laws of planetary motion: planets move on ellipses with the sun at the focus; the line from the sun to any planet sweeps over equal areas in equal times; and the squares of the periods (the times it takes the various planets to go around their orbits) are proportional to the cubes of the major diameters of the planets' orbits.

It is usual to say that Newton's laws explain Kepler's. But historically Newton's law of gravitation was deduced from Kepler's laws of planetary motion. Edmund Halley, Christopher Wren, and Robert Hooke all used Kepler's relation between the squares of the periods and the cubes of the diameters (taking the orbits as circles) to deduce an inverse square law of gravitation, and then Newton extended the argument to elliptical orbits. Today, of course, when you study mechanics you learn to deduce Kepler's laws from Newton's laws, not vice versa. We have a deep sense that Newton's laws are more fundamental than Kepler's laws, and it is in that sense that Newton's laws explain Kepler's laws rather than the other way around. But it's not easy to put a precise meaning to the idea that one physical principle is more fundamental than another.

It is tempting to say that more fundamental means more comprehensive. Perhaps the best-known attempt to capture the meaning that scientists give to explanation was that of Carl Hempel. In his 1948 article written with Paul Oppenheim,[3] he remarked that "the explanation of a general regularity consists in subsuming it under another more comprehensive regularity, under a more general law." But this doesn't remove the difficulty. One might say for instance that Newton's laws govern not only the motions of planets but also the tides on Earth, the falling of fruits from trees, and so on, while Kepler's laws deal with the more limited context of planetary motions. But that isn't strictly true. Kepler's laws, to the extent that classical mechanics applies at all, also govern the motion of electrons around the nucleus in atoms, where gravity is irrelevant. So there is a sense in which Kepler's laws have a generality that Newton's laws don't have. Yet it would feel absurd to say that Kepler's laws explain Newton's, while everyone (except perhaps a philosophical purist) is comfortable with the statement that Newton's laws explain Kepler's.

This example of Newton's and Kepler's laws is a bit artificial, because there is no real doubt about which is the explanation of the other. In other cases the question of what explains what is more difficult, and more important. Here is an example. When quantum mechanics is applied to Einstein's General Theory of Relativity one finds that the energy and momentum in a gravitational field come in bundles known as gravitons, particles that have zero mass, like the particle of light, the photon, but have a spin equal to two (that is, twice the spin of the photon). On the other hand, it has been shown that any particle whose mass is zero and whose spin is equal to two will behave just the way that gravitons do in General Relativity, and that the exchange of these gravitons will produce just the gravitational effects that are predicted by General Relativity. Further, it is a general prediction of string theory that there must exist particles of mass zero and spin two. So is the existence of the graviton explained by the General Theory of Relativity, or is the General

3. Carl Hempel and Paul Oppenheim, "Studies in the Logic of Confirmation," *Philosophy of Science* 15, no. 135 (1948), pp. 135–175; reprinted with some changes in *Aspects of Scientific Explanation and Other Essays in the Philosophy of Science* (Free Press, 1965).

Theory of Relativity explained by the existence of the graviton? We don't know. On the answer to this question hinges a choice of our vision of the future of physics—will it be based on space-time geometry, as in General Relativity, or on some theory like string theory that predicts the existence of gravitons?

The idea of explanation as deduction also runs into trouble when we consider physical principles that seem to transcend the principles from which they have been deduced. This is especially true of thermodynamics, the science of heat and temperature and entropy. After the laws of thermodynamics had been formulated in the nineteenth century, Ludwig Boltzmann succeeded in deducing these laws from statistical mechanics, the physics of macroscopic samples of matter that are composed of very large numbers of individual molecules. Boltzmann's explanation of thermodynamics in terms of statistical mechanics became widely accepted, even though it was resisted by Max Planck, Ernst Zermelo, and a few other physicists who held on to the older view of the laws of thermodynamics as freestanding physical principles, as fundamental as any others. But then the work of Jacob Bekenstein and Stephen Hawking in the twentieth century showed that thermodynamics also applies to black holes, and not because they are composed of many molecules, but simply because they have a surface from which no particle or light ray can ever emerge.[4] So thermodynamics seems to transcend the statistical mechanics of many-body systems from which it was originally deduced.

Nevertheless, I would argue that there is a sense in which the laws of thermodynamics are not as fundamental as the principles of General Relativity or the Standard Model of elementary particles. It is important here to distinguish two different aspects of thermodynamics—as a mathematical formalism, and as a state-

4. Added note: This isn't quite true. For thermodynamics to apply to a black hole, it is also necessary that the surface be sufficiently large—its dimensions must be much larger than the Planck length, about 10^{-33} centimeters, the scale of distances below which quantum fluctuations in the gravitational field become appreciable.

ment about the real world. On one hand, thermodynamics is a formal system that allows us to deduce interesting consequences from a few simple laws, wherever those laws apply. The laws apply to black holes, they apply to steam boilers, and to many other systems. But they don't apply everywhere. Thermodynamics would have no meaning if applied to a single atom. To find out whether the laws of thermodynamics apply to a particular physical system, you have to ask whether the laws of thermodynamics can be deduced from what you know about that system. Sometimes they can, sometimes they can't. Thermodynamics by itself is never the explanation of anything—you always have to ask why thermodynamics applies to whatever system you are studying, and you do this by deducing the laws of thermodynamics from whatever more fundamental principles happen to be relevant to that system.

In this respect, I don't see much difference between thermodynamics and Euclidean geometry. After all, Euclidean geometry applies in an astonishing variety of contexts. If three people agree that each one will measure the angle between the lines of sight to the other two, and then they get together and add up those angles, the sum will be 180 degrees. And you will get the same 180-degree result for the sum of the angles of a triangle made of steel bars or of pencil lines on a piece of paper. So it may seem that geometry is more fundamental than optics or mechanics. But Euclidean geometry is a formal system of inference based on postulates that may or may not apply in a given situation. As we learned from Einstein's General Theory of Relativity, the Euclidean system does not apply in gravitational fields, though it is a very good approximation in the relatively weak gravitational field of the Earth in which it was developed by Euclid and his predecessors. When we use Euclidean geometry to explain anything in nature we are tacitly relying on General Relativity to explain why Euclidean geometry applies in the case at hand.

In talking about deduction, we run into another problem: Who is it that is doing the deducing? We often say that something is explained by something else without our actually being able to deduce it. For example, after the development of quantum mechanics in the mid-1920s, when it became possible for the first time to calculate in a clear and understandable way the binding energy and spectrum of the hydrogen molecule, many physicists immediately concluded that all of chemistry is explained by quantum mechanics and the principle of electrostatic attraction and repulsion among electrons and atomic nuclei. Physicists like Paul Dirac proclaimed that now all of chemistry had become understood. But they had not yet succeeded in deducing the chemical properties of any molecules except the simplest hydrogen molecule. Physicists were sure that all these chemical properties were consequences of the laws of quantum mechanics as applied to atomic nuclei and electrons.

Experience has borne this out; we now can in fact deduce the properties of fairly complicated molecules—not molecules as complicated as proteins or DNA, but still some fairly impressive organic molecules—by doing complicated calculations using quantum mechanics and the principle of electrostatic attraction and repulsion. Almost any physicist would say that chemistry is explained by quantum mechanics and the simple properties of electrons and atomic nuclei. But chemical phenomena will never be entirely calculated in this way, and so chemistry persists as a separate discipline. Chemists do not call themselves physicists; they have different journals and different skills from physicists. It's difficult to deal with complicated molecules by the methods of quantum mechanics, but still we know that physics explains why chemicals are the way they are. The explanation is not in our books, it's not in our scientific articles, it's in nature; it is that the laws of physics require chemicals to behave the way they do.

Similar remarks apply to other areas of physical science. As part of the Standard Model, we have a well-verified theory of the strong nuclear force—the force that binds together both the protons and neutrons in the atomic nucleus and the quarks that make up the protons and neutrons—known as quantum chromodynamics, which we believe explains why the proton (and neutron) mass is what it is. The proton mass is almost entirely produced by the strong forces that the quarks inside the proton exert on one another. It is not that we can actually calculate the proton mass; I'm not even sure we have a good algorithm for doing the calculation, but there is no sense of mystery about the mass of the proton. We feel we know why it is what it is, not in the sense that we have calculated it or even can calculate it, but in the sense that quantum chromodynamics can calculate it—the value of the proton mass is *entailed* by quantum chromodynamics, even though we don't know how to do the calculation.[5]

It can be very important to recognize that something has been explained, even in this limited sense, because it can give us a strategic sense of what problems to work on. If you want to work on calculating the proton mass, go ahead, more power to you. It would be a lovely show of calculational ability, but it would not advance our understanding of the laws of nature, because we already understand the strong nuclear force well enough to know that no new laws of nature will be needed in this calculation.

Another problem with explanation as deduction: in some cases we can deduce something without explaining it. That may sound really peculiar, but consider the following little story. When physicists started to take the big bang cosmology seriously, one of the things they did was to calculate the production of light elements in the first few minutes of the expanding universe. The way this was done was to write down all the equations that govern the rates at which various nuclear reactions took place. The rate of change of the quantity (or "abundance," as physicists say) of any one nuclear species is equal to a sum of terms, each term proportional to the abundance of that or some other nuclear species. In this way you develop a large set of linked differential equations, and then you put them on a computer that produces a numerical solution.

When these equations were solved in the mid-1960s by James Peebles and then by Robert Wagoner, William Fowler, and Fred

---

5. Added note: At the time this was written, theorists were exploring a possible method of approximate calculation, called lattice quantization, in which the space-time continuum is replaced with a lattice of closely packed discrete points. Since then this method had been developed so far that it has become possible to calculate the mass of the proton and other particles, obtaining good agreement with measured values.

Hoyle, it was found that after the first few minutes about a quarter of the mass of the universe was left in the form of helium, and almost all the rest was hydrogen, with other elements present only in tiny quantities. These calculations also revealed certain regularities. For instance, if you put something in the theory to speed up the expansion, as for instance by adding additional species of neutrinos, you would find that more helium would be produced. This is somewhat counterintuitive—you might think that speeding up the expansion of the universe would leave less time for the nuclear reactions that produce helium, but in fact the calculations showed that it increased the amount of helium produced.

The explanation is not difficult, though it can't easily be seen in the computer printout. While the universe was expanding and cooling in the first few minutes, nuclear reactions were occurring that built up complex nuclei from the primordial protons and neutrons, but because the density of matter was relatively low, these reactions could occur only sequentially, first by combining some protons and neutrons to make the nucleus of heavy hydrogen, the deuteron, and then by combining deuterons with protons or neutrons or other deuterons to make heavier nuclei like helium. However, deuterons are very fragile; they're relatively weakly bound, so essentially no deuterons were produced until the temperature had dropped to about a billion degrees, at the end of the first three minutes. During all this time neutrons were changing into protons, just as free neutrons do in our laboratories today.

When the temperature dropped to a billion degrees, and it became cold enough for deuterons to hold together, then all of the neutrons that were still left were rapidly gobbled up into deuterons, and the deuterons then into helium, a particularly stable nucleus. It takes two neutrons as well as two protons to make a helium nucleus, so the number of helium nuclei produced at that time was just half the number of remaining neutrons. Therefore the crucial thing that determines the amount of helium produced in the early universe is how many of the neutrons decayed before the temperature dropped to a billion degrees. The faster the expansion went, the earlier the temperature dropped to a billion degrees, so the less time the neutrons had to decay, so the more of them were left, and so the more helium was produced. That's the explanation of what was found in the computer calculations; but the explanation was not to be found in the computer-generated graphs showing the abundance in relation to the speed of expansion.

Further, although I have said that physicists are only interested in explaining general principles, it is not so clear what is a principle and what is a mere accident. Sometimes what we think is a fundamental law of nature is just an accident. Kepler again provides an example. He is known today chiefly for his famous three laws of planetary motion, but when he was a young man he tried also to explain the diameters of the orbits of the planets by a complicated geometric construction involving regular polyhedra nested between spheres on which the planets ride. Today we smile at this because we know that the distances of the planets from the Sun reflect accidents that occurred as the solar system happened to be formed. We wouldn't try to explain the diameters of the planetary orbits by deducing them from some fundamental law.

In a sense, however, there is a kind of approximate statistical explanation for the distance of the Earth from the Sun.[6] If you ask why the Earth is about a hundred million miles from the Sun, as

6. Professor R. J. Hankinson of the University of Texas has directed my attention to Galen for an early example of this "explanation." Of course, writing fourteen hundred years before Copernicus, Galen was concerned to explain the position of the Sun rather than that of Earth. In "On the Usefulness of the Parts of the Body" he compared his explanation of the Sun's position to the explanation of the position of the human foot at the end of the leg—both Sun and foot are placed by the creator where they would do the most good. Although these explanations are teleological in a way that has been abandoned by modern science, Galen's analogy was better than he could have realized. Just as Earth is one of a vast number of planets, whose distances from their stars is largely a matter of chance, so the position of the foot is the outcome of a vast number of chance mutations in the evolution of our vertebrate ancestors. An organism produced by a chain of chance mutations that put its feet in its mouth would not survive to pass its genes on to its descendants, just as a planet that by chance condensed too close to or too far from its star would not be the home of philosophers.

opposed, say, to two hundred million or fifty million miles, or even further, or even closer, one answer would be that if the Earth were much closer to the Sun, then it would be too hot for us, and if it were any further from the Sun, then it would be too cold for us. As it stands, that's a pretty silly explanation, because we know that there was no advance knowledge of human beings in the formation of the solar system. But there is a sense in which that explanation is not so silly. There are countless planets in the universe, so that even if only a tiny fraction are the right distance from their stars and have the right mass and chemical composition and so on to allow life to evolve, it should be no surprise that creatures that inquire into the distance of their planet from its star would find that they live on one of the planets in this tiny fraction.

This kind of explanation is known as anthropic, and as you can see, it does not offer a terribly useful insight into the physics of the solar system. But anthropic arguments may become very important when applied to what we usually call the universe. Cosmologists increasingly speculate that just as the Earth is just one of many planets, so also our big bang, the great expansion of the universe in which we live, may be just one of many bangs that go off sporadically here and there in a much larger "multiverse." They further speculate that in these many different big bangs some of the supposed constants of nature take different values, and perhaps even some of what we now call the laws of nature take different forms. In this case, the question why the laws of nature that we discover and the constants of nature that we measure are what they are would have a rough teleological explanation—that it is only with this sort of big bang that there would be anyone to ask the question.

I certainly hope that we will not be driven to this sort of reasoning, and that we will discover a unique set of laws of nature that explain why all the constants of nature are what they are. But we have to keep in mind the possibility that what we now call the laws of nature and the constants of nature are accidental features of the big bang in which we happen to find ourselves, though constrained (as is the distance of the Earth from the Sun) by the requirement that they have to be in a range that allows the appearance of beings that can ask why they are what they are.

Conversely, it is also possible that a class of phenomena may be regarded as mere accidents when in fact they are manifestations of fundamental physical principles. I think this may be the answer to a historical question that has puzzled me for many years. Why was Aristotle (and many other natural philosophers, notably Descartes) satisfied with a theory of motion that did not provide any way of predicting where a projectile or other falling body would be at any moment during its flight, a prediction of the sort that Newton's laws do provide? According to Aristotle, substances tend to move to their natural positions—the natural position of earth is downward, the natural position of fire is upward, and water and air are naturally somewhere in between—but Aristotle did not try to say how fast a bit of earth drops downward or a spark flies upward. I am not asking why Aristotle had not discovered Newton's laws—obviously someone had to be the first to discover these laws, and the prize happened to go to Newton. What puzzles me is why Aristotle expressed no dissatisfaction that he had not learned how to calculate the positions of projectiles at each moment along their paths. He did not seem to realize that this was a problem that anyone ought to solve.

I suspect that this was because Aristotle implicitly assumed that the rates at which the elements move to their natural places are mere accidents, that there is nothing interesting that you could say about them (except for his notion that the rate of free fall is proportional to the weight), that the only really interesting things were questions of equilibrium—where objects will come to rest. This may have reflected a widespread disdain for change on the part of the Hellenic philosophers, as shown for instance in the work of Parmenides, which was admired by Aristotle's teacher Plato. Of course Aristotle was wrong about this, but if you imagine yourself in his times, you can see how far from obvious it would have been that motion is governed by precise mathematical rules that might be discovered. As far as I know, this was not understood until Galileo began to measure how long it took balls to roll various distances down an inclined plane.[7] It is one of the great tasks of science to learn what are accidents and what are principles, and about this we cannot always know in advance.

So now that I have deconstructed the words "fundamental," "deduce," and "principle," is anything left of my proposal that in physics we say that we explain a principle when we deduce it from a more fundamental principle? Yes, I think there is, but only within a historical context, a vision of the future of science. We have been steadily moving toward a satisfying picture of the world. We hope that in the future we will have achieved an understanding of all the regularities that we see in nature, based on a few simple principles, laws of nature, from which all other regularities can be deduced. These laws will be the explanation of whatever principles (such as, for instance, the rules of the Standard Model or of General Relativity) can be deduced directly from them, and those directly deduced principles will be the explanations of whatever principles can be deduced from them, and so on. Only when we have this final theory will we know for sure what is a principle and what an accident, what facts about nature are entailed by what principles, and which are the fundamental principles and which are the less fundamental principles that they explain.

I have now done the best I can to say whether science can explain anything, so let me take up the question whether science can explain everything. Clearly not. There certainly always will be accidents that no one will explain, not because they could not be explained if we knew all the precise conditions that led up to them, but because we never will know all these conditions. There are questions like why the genetic code is precisely what it is or why a comet happened to hit the earth sixty-five million years ago in just

---

7. Added note: The distance traveled under conditions of uniform acceleration was worked out in the fourteenth century, but as far as I know this result was not applied then to falling bodies.

the place it did rather than somewhere else that will probably remain forever outside our grasp. We cannot explain, for example, why John Wilkes Booth's bullet killed Lincoln in 1865 while Giuseppe Zangara's shots missed Franklin Roosevelt in 1933. We might have a partial explanation if we had evidence that Zangara's arm was jostled just as he pulled the trigger; but, as it happens, we don't. All such information is lost in the mists of time; events depend on accidents that we can never recover. We can perhaps try to explain them statistically; for example, you might consider a theory that southern actors in the nineteenth century tended to be good shots while Italian anarchists in the twentieth century tended to be poor shots, but when you only have a few singular pieces of information, it's very difficult to make even statistical inferences. Physicists try to explain just those things that are not dependent on accidents, but in the real world most of what we try to understand does depend on accidents.

Further, science can never explain any moral principle. There seems to be an unbridgeable gulf between "is" questions and "ought" questions. We can perhaps explain why people think they should do things, or why the human race has evolved to feel that certain things should be done and other things should not, but it remains open to us to transcend these biologically based moral rules. It may be, for example, that our species has evolved in such a way that men and women play different roles—men hunt and fight, while women give birth and care for children—but we can try to work toward a society in which every sort of work is as open to women as it is to men. The moral postulates that tell us whether we should or should not do so cannot be deduced from our scientific knowledge.

There are also limitations on the certainty of our explanations. I don't think we'll ever be certain about any of them. Just as there are deep mathematical theorems that show the impossibility of proving that arithmetic is consistent, it seems likely that we will never be able to prove that the most fundamental laws of nature are mathematically consistent. Well, that doesn't worry me, because even if we knew that the laws of nature are mathematically consistent, we still wouldn't be certain that they are true. You give up worrying about certainty when you make that turn in your career that makes you a physicist rather than a mathematician.

Finally, it seems clear that we will never be able to explain our most fundamental scientific principles. (Maybe this is why some people say that science does not provide explanations, but by this reasoning nothing else does either.) I think that in the end we will come to a set of simple universal laws of nature, laws that we cannot explain. The only kind of explanation I can imagine (if we are not just going to find a deeper set of laws, which would then just push the question farther back) would be to show that mathematical consistency requires these laws. But this is clearly impossible, because we can already imagine sets of laws of nature that, as far as we can tell, are completely consistent mathematically but that do not describe nature as we observe it.

For example, if you take the Standard Model of elementary particles and just throw away everything except the strong nuclear forces and the particles on which they act, the quarks and the gluons, you are left with the theory known as quantum chromodynamics. It seems that quantum chromodynamics is mathematically self-consistent, but it describes an impoverished universe in which there are only nuclear particles—there are no atoms, there are no people. If you give up quantum mechanics and relativity, then you can make up a huge variety of other logically consistent laws of nature, like Newton's laws describing a few particles endlessly orbiting each other in accordance with these laws, with nothing else in the universe, and nothing new ever happening. These are logically consistent theories, but they are all impoverished. Perhaps our best hope for a final explanation is to discover a set of final laws of nature and show that this is the only logically consistent rich theory, rich enough for example to allow for the existence of ourselves. This may happen in a century or two, and if it does then I think that physicists will be at the extreme limits of their power of explanation.

This essay evoked two interesting comments from readers of the *New York Review of Books*. David Lowenthal, a professor of geography at University College, London, thought that I had slighted historians by saying that they, unlike physicists, are interested in the causes of individual events, such as the French Revolution. I explained in a published reply that I did not mean to imply that historians have no interest in regularities in history, but rather that historians often find individual events also interesting in their own right, not just as clues to regularities, while physicists, unlike historians (or biologists or meteorologists), are at bottom interested in nothing but regularities. In another letter Rabbi Everett Gendler, who as it happened had taught our daughter at Andover, took me to task for saying that "we know that there was no advance knowledge of human beings in the formation of the solar system." I answered that he was right if by "know" is meant "know with certainty" or "know by direct observation." I explained that what I should have said is that nothing in our best scientific theories of the origin of the solar system suggests that advance knowledge of human beings was involved. I added that it has been an essential element in the success of science to distinguish those problems that are or are not illuminated by taking human beings into account.

3

Peace at Last in the Science Wars

The "science wars" of the 1990s were debates between an assortment of sociologists, historians, and cultural critics, who emphasized the social and cultural influences on scientific research, and scientists, who instead viewed their work as primarily an objective search for truth. By 2000 both parties had discovered a good deal of common ground, and the argument had largely quieted down. A Cal Tech chemist, Jay Labinger, and a sociologist at Cardiff University, Harry Collins, had the idea of bringing together essays from various scholars and scientists who had participated in the science wars, to clarify any remaining disagreements. The essays were collected in a book, *The One Culture?* published in 2001 by the University of Chicago Press. (The title of the book was meant to remind the reader of C. P. Snow's celebrated 1959 Rede Lecture, "The Two Cultures," though Snow had been complaining about the fact that humanistic culture was ignoring science, while the science wars arose because the humanists were giving science what scientists felt was the wrong sort of attention.) *The One Culture?* was divided into three parts: Positions, Commentaries, and Rebuttals. When asked to contribute, to state my position I submitted a merged version of two articles, "Physics and History" and "The Non-revolution of Thomas Kuhn," both of which had appeared in my previous essay collection, *Facing Up*. As a commentary, I submitted the article that follows here. I also contributed a brief rebuttal to comments made by Harry Collins, Peter Dear, and Trevor Pinch, but this rebuttal would only be of interest to someone who had read their remarks (if, indeed, even to such a reader) and so is not included in this collection.

In reading the other essays in this book I was disappointed to see how much there was in them with which I could agree. The essays by sociologists, historians, and philosophers did not criticize the way that science is done, any more than the essays by scientists objected to the study of science as a social enterprise. We seem to be hurtling toward a general reconciliation. But perhaps it is still not too late to draw back from the brink of peace.

It seems to me that there is at least one important issue left to be argued. Although scientists recognize that their theories often bear the stamp of the social environment in which they are formulated, we like to think of this as an impurity, some slag left amid the metal, which we hope eventually to eliminate.[1] We have felt the powerful attraction that true theories exert on our thinking, an attraction that seems to have little to do with the social setting of our research. We wonder why some historians of science like Harry Collins are not interested in describing this process, the often slow and uncertain progress of physical theories toward an ultimate culture-free form that is the way it is because this is the way the world is.

In a 1996 article[2] I tried to express this view of science by saying that when we make contact with beings from another planet we will find that they have discovered the same laws of physical science as we have. In his essay in *The One Culture?* Michael Lynch[3] very properly caught me up on this and pointed out that the scientific works of the intelligent creatures who inhabited Earth until a few centuries ago did not resemble the theories we believe in today. But this just underlines the point that I am trying to make here, that it is not necessarily the state of scientific theory at the moment that is culture-free, but its asymptotic form, the

---

1. Well, at least physicists. Or at least some physicists. Or maybe just me.

2. "Sokal's Hoax," *New York Review of Books* 43, no. 13 (August 8, 1996): 11–15; reprinted in *Facing Up*.

3. "Is a Science Peace Process Necessary?" in *The One Culture*, ed. J. A. Labinger and H. Collins (University of Chicago Press 2001), 48–60.

form which it ultimately tends. As I understand it, most sociologists of science either deny the existence of this asymptotic limit or choose to ignore it.

In arguing about this, scientists are likely to mutter about "truth" and "reality." We talk about the truth of our theories and how they correspond to something real. This is a dangerous business; philosophers have been arguing for centuries about what is meant by truth or reality. Ian Hacking[4] calls these "elevator words," by which he means that they are used to elevate the importance of what we are talking about, without really being of much help in settling anything. I have myself expressed the opinion that when we say that a thing is real we are simply expressing a sort of respect.[5] In a more negative spirit, Thomas Kuhn has said that "no sense can be made of the notion of reality as it has ordinarily functioned in the philosophy of science."[6] So it is not surprising that both Michael Lynch[7] and Steven Shapin[8] in their essays in *The One Culture?* and Richard Rorty elsewhere[9] took issue with my statement that "for me as a physicist the laws of nature are real in the same sense (whatever that is) as the rocks on the ground."

Of course in saying this I did not think that I had solved the ancient ontological problems surrounding the concept of reality. This is why I inserted the parenthetic clause "whatever that is" in speaking of the sense in which rocks are real. Maybe I should have been more explicit in my modesty, which Shapin took as nervousness.

4. Ian Hacking, *The Social Construction of What?* (Harvard University Press, 1999).

5. *Dreams of a Final Theory* (Pantheon, 1992).

6. "The Trouble with the Historical Philosophy of Science," Rothschild Distinguished Lecture in the Department of the History of Science, Harvard University, November 19, 1991.

7. See *supra* note 3.

8. "How to Be Unscientific," in *The One Culture,* ed. J. A. Labinger and H. Collins (University of Chicago Press 2001): 99–115.

9. "Thomas Kuhn, Rocks, and the Laws of Physics," *Common Knowledge* 6, no. 1 (1997): 6–16.

On the other hand, I do want to defend the use of words like "truth" and "reality" in discussing the history and sociology of science. Everyone uses these words in everyday life: the monster in my dream last night was not real, the deer on the lawn this morning is real, and that's the truth. We do not have a precise conception of what we mean by these words, any more than we have a precise conception of what we mean by other words like "cause" or "love" or "beauty," but they are still useful to us. When we say that something is real we intend to convey that our experience of it has one or more of those properties that give us the impression that something has an independent existence: maybe other people can experience it, maybe it doesn't change when our mood changes; maybe it is something about which it is possible to make mistakes; maybe it doesn't go away when we look at it more closely. Individual rocks are eminently real in this sense, though the category "rocks" may not be. My comparison of the laws of nature to rocks was not a philosophical argument, but rather a personal report, that my experience of the laws of nature in my work as a physicist has the same qualities that in the case of rocks make me say that rocks are real.

This raises a larger issue. As shown by our common use of words like "real" and "true," we all adopt a working philosophy in our everyday lives that can be called naïve realism. As far as I know, no one has shown why we should abandon naive realism when talking of the history and sociology of science. Philosophers may be able to help us to sharpen the way we understand words like "real" and "true" and "cause," but they have no business telling us not to use them.

4

The Future of Science,
and the Universe

To celebrate the new century, the New York Public Library joined with the *New York Review of Books* in 2001 to sponsor a series of lectures entitled "Futures: Bright, Dim, and Otherwise." I gave one of these talks at the Library in January 2001. The following essay was based on my talk, and published in the *New York Review* in November 2001.

   In discussing the future of science, I went over some of the same ground as in my book *Dreams of a Final Theory*. But in talking about the future of the universe, I did bring in one piece of current news: Only two years earlier astronomers had found evidence that the expansion of the universe, long thought to be slowing down because of the gravitational attraction of galaxies for each other, was instead speeding up. This is attributed to a "dark energy," an energy resident in space itself, rather than in any particles of matter and radiation. The dark energy is the subject of the next essay in this collection.

In the program for this lecture series at the New York Public Library I saw one vision of the future: Raymond Loewy's conception of an airliner, as exhibited at the 1939 World's Fair in New York. I was there at the 1939 World's Fair, but I don't remember Raymond Loewy's design. I was very young. What I best remember are the fountains lit up by colored lights. Also, I remember that a dairy company was giving out tiny free ice cream cones. With the depression still going on, free ice cream was a memorable experience. Whatever predictions of future technology were made at the World's Fair did not leave much of an impression on me.

It was no great loss. Aside perhaps from the vision of modern superhighways in the General Motors pavilion, the World's Fair did not score great successes in its predictions of future technology. The illustration of Loewy's design for an airliner of the future doesn't look at all like passenger aircraft today. It shows eight engines, and a fuselage resembling a diesel locomotive. I didn't know it in 1939, but Raymond Loewy had in fact designed diesel locomotives for the Pennsylvania Railroad in the 1930s, giving them a futuristic "streamlined" look without actually paying much attention to principles of aerodynamics. He could get away with this with diesel locomotives, but not with airplanes. But predicting future technology is very difficult even if you don't ignore the laws of physics. You might better spend your time admiring fountains of colored water.

My subject here is not the future of technology or other applications of science, but the future of science itself. Here we can make a prediction with fair confidence—that sooner or later we shall discover the physical principles that govern all natural phenomena.

We already have a theory that encompasses all the particles out of which we and our surroundings are made, and, except for gravitation, all the forces that act on them. This theory, known as the Standard Model, is expressed in the mathematical formalism of quantum mechanics, which seems to be a universal basis for the laws of physics. But the Standard Model has too many arbitrary elements, like the masses that have to be assigned to the various elementary particles. We also have a theory of gravitation—the General Theory of Relativity—which predated quantum mechanics. This theory can even be interpreted quantum mechanically. The trouble is that our present quantum theory of gravitation only provides approximations whose validity is limited to processes at sufficiently low energies and large distances. What we do not yet have is a quantum mechanical theory of unlimited validity that encompasses all particles and forces. We are like the plebeians of Rome at the time when the Twelve Tables were still kept secret, not knowing the laws by which we are governed. More specifically, as I will come to later, our ignorance of the final laws of nature makes us uncertain in our predictions of the future of the universe.

There is now a strong suspicion that the final theory will be something like today's string theories, which you can read about in the popular book by Brian Greene.[1] To describe them briefly (which is hardly possible): In a string theory each elementary particle, whether it is an electron or a neutrino or a quark or whatever, is in reality a string, a tiny one-dimensional entity that vibrates as it zips through space, with all the different types of elementary particles corresponding to different modes of vibration of the string. In essentially all string theories it turns out that one of these particle types is the graviton, the massless particle responsible for gravitational forces in the quantum theory of gravitation. So string theory automatically brings gravitation into the picture, along with the other forces.

For some time it seemed that there were five different mathematically consistent string theories, which would be a depressing result, because no one had any insight on why nature would be described by one of these theories rather than one of the other four. To avoid inconsistencies, all of these string theories were formulated in ten space-time dimensions (nine space dimensions plus the dimension of time), which was even more depressing because there is pretty good evidence that we don't live in ten dimensions. But it has recently been found (by Edward Witten) that these five different string theories (as well as one nonstring theory in eleven space-time dimensions) are all just different phases of one underlying theory that unfortunately we do not yet know. There are also some reasons to hope that our successful Standard Model of particles and forces in four space-time dimensions—length, width,

---

1. Brian Greene, *The Elegant Universe: Superstrings, Hidden Dimensions, and the Quest for the Ultimate Theory* (Norton, 1999).

depth, and time—will be found to be another phase, the one in which we live, of this underlying theory.

It's a little bit like saying that diamond and graphite are phases of the same material, carbon. If the only thing anyone knew about carbon came from observations of the diamonds in jewelry and the graphite in pencils, it might be very hard to learn that these are phases of the same substance. One could see that there were some things diamond and graphite have in common—the way they interact with neutrons from a nuclear reactor, for instance. But if you didn't know about the carbon atom, it would be extraordinarily difficult to find a unified theory of diamond and graphite that revealed they were really just different forms of carbon. Of course, this is just an analogy. I am not talking here about different phases of a substance but about different phases of a physical theory, phases characterized by different numbers of dimensions in space and time, among other things. The hope is that these different theories are just approximations to different solutions of the equations of the underlying theory, more or less as diamond and graphite can be understood as different solutions of the equations governing carbon atoms.

How are we going to find the theory that underlies all these different versions of string theory, plus the theory that describes observed phenomena in our own four-dimensional space-time? The answer is—with difficulty. The time scale for this discovery may be anything from hours to centuries. Tomorrow, when I look at the web site that I check every morning to see what is new in physics, I may find an article by some previously unknown graduate student, laying it all out. Then again, it may not happen in this century. But I think it will happen, and when it does it will end a certain chapter in the history of science: the search for the fundamental principles that underlie everything.[2]

It is very likely that the final theory we find in this way will be quite simple, in the sense of being based on only a few fundamen-

---

2. I have written about this at greater length in *Dreams of a Final Theory* (Pantheon, 1993).

tal principles. It will probably also be very fragile, in the sense that it will not be possible to make any small change in the theory without its becoming logically inconsistent. We have already gone a great way in this direction. For instance, quantum mechanics, which was developed in the mid-1920s, has survived essentially unchanged to the present day. If you try to think of theories that are similar to quantum mechanics but only a little different, you find they always involve logical absurdities like negative probabilities or causes following effects. When you combine quantum mechanics with relativity, the fragility increases. You find that if you do not construct a relativistic quantum theory carefully, then when you ask a reasonable question, like asking for the rate of a certain reaction, you get an unreasonable answer—the rate is infinite.

Only certain limited classes of theories avoid these nonsensical infinities.[3] This fragility is a good thing, because it can go some way toward telling us why the laws of nature are what they are. We can hope that with its greater fragility, a final theory will not involve any free parameters, like the particle masses in the Standard Model, whose numerical values have to be taken from experiment without our understanding why these numbers are what they are. Fragility also gives our theories much of their beauty, in the same way that a Chopin waltz gains beauty from our sense that no note in it could be changed.

Achieving a final theory will be a great achievement, but it will not be entirely satisfactory, because although we will then know why the theory is not *slightly* different from other consistent theories, we will never know why the theory is not something *com-*

---

3. Added note: Strictly speaking, this is only true if we insist on theories that describe nature at all distance scales. There is a large class of what are called effective field theories, which provide excellent approximations at accessible distance scales but lose their validity at very small scales. It is thought that today's successful field theories, such as General Relativity and the Standard Model of elementary particle physics, are effective field theories. The large-distance approximations to effective field theories are subject to strict limitations, but not as strict as those that must be satisfied by theories that describe nature at all distance scales.

pletely different. For instance, if you are willing to abandon either quantum mechanics or relativity altogether, then you can construct any number of theories that are logically consistent, but that do not describe the real world.

Another limitation: This theory, although I call it a final theory, will not be the end of the road for science. It will not be what is sometimes called a theory of everything, a theory that solves all scientific problems. We already have many scientific problems that will not be illuminated at all. One of them, to take a problem within physics, is to understand the flow of a fluid when it becomes turbulent. This problem has faced us for one hundred years; it still defies solution and may go on resisting solution well after the success of the final theory of elementary particles, because we already understand all we need to know about the fundamental principles governing fluids. We just don't know how to deal with the complicated kind of fluid flow in which eddies are carried by larger eddies carried by larger eddies and so on, the characteristic feature of fully developed turbulence. As with many of the most interesting problems of physics, computers are only of limited help in understanding the turbulence of fluids, because they just tell us what happens in a variety of special circumstances, which we could also learn from experiment. What we would really like to understand are the universal properties of fully developed turbulence under all circumstances.

And, of course, a final theory of physics will not be of much use to our friends in biology. Very likely developments in biology have had and will go on having the largest impact on human culture. One of the greatest moments in the history of human thought was the discovery by Darwin and Wallace in the nineteenth century that no divine intervention or "life force" is needed to explain the evolution of species. Life is not governed by independent fundamental biological laws—it can be described as the effects of physics and chemistry worked out over billions of years of accidents. It is not so long ago that many people's religious beliefs were based on the argument from design, the argument that the wonderful characteristics of living things could not possibly arise without a divine plan. Lytton Strachey tells how Cardinal Manning came to his faith in just this way. Now that we understand how evolution can occur through the natural selection of random mutations, the argument from design has lost its force for anyone with a reasonable understanding of biological science.

The big challenge ahead in biology is to understand behavior, which seems to be far more difficult than understanding other aspects of life. I am far from my field of expertise in talking about behavior, but perhaps I can be forgiven for saying that I don't think that it is an insuperable scientific problem. Very likely this problem will not first be solved for human beings, but for a very well-studied nematode worm, whose full name I can neither spell nor pronounce, but is abbreviated as *C. elegans*. This worm is one of the classic animals that biologists study, like the mouse and the fruit fly. The worm had its entire nervous system mapped out some years ago, but we don't understand its behavior because we don't know the program that governs its behavior. We are in a position like that of an industrial spy, who has bought a personal computer, and can map out the connections of every transistor on its central processing chip and locate every magnetized dot on its hard drive, and has looked over the shoulders of people using the Windows operating system on the computer, but has not been able to figure out how the transistors and magnetized dots make the operating system work. (I don't know why anyone would want to steal that particular operating system, but we can agree that it's a very difficult problem.)

This problem has not been solved, but it doesn't seem insoluble for *C. elegans,* and eventually it will be solved for human beings. I don't mean that we will necessarily solve the problems about consciousness that bother philosophers. (How can I know that you and I perceive the same thing when we both see the color red?) But I do think that we will have an understanding of behavior, and of whatever aspects of consciousness are necessary to understand behavior. It won't be a completely predictive theory, like our theory of eclipses, because behavior is too complicated for that, but more like our theory of weather. We can't always predict the weather, but we know pretty well how the weather works.

Implicit in all this is a conservative assumption, that science will continue indefinitely in the path laid out by Galileo and Newton: the discovery of increasingly comprehensive mathematical laws that will increasingly be shown to account, aside from historical accidents, for all phenomena, biological as well as physical. I can't be sure that this is correct. Just as medieval Aristotelians could hardly imagine science in the style of Newton, so science in the future may take a turn that we cannot now imagine. But I see not the slightest advance sign of such a change.

The developments in physics that I have described have already illuminated the future of the universe. In fact, unlike the problem of predicting the future of elections or the stock market, it's much easier to make predictions about the future of the universe than calculations of what must have happened in the past. The universe is expanding and cooling, so when you look back in time you have to consider an era of enormous density and extremely high temperature, densities and temperatures at which our present theories become inapplicable. Without a final, universally applicable theory, we cannot penetrate theoretically to the first tiny fraction of a second and understand what happened at the very beginning of the big bang. But looking into the future, we see that the universe will expand, getting less dense and colder, and so it becomes easier and easier to describe—at least for a while.

What I mean by the expansion of the universe needs explanation, because misunderstandings about it keep coming up. When astronomers say the universe is expanding, they don't mean that space is expanding, although sometimes some of us are guilty of putting it that way. Objects that are bound together, like galaxies and solar systems and tape measures, are not expanding. In fact, if tape measures and other standards of length and everything else were expanding the same way that the universe was expanding, how would you know that anything was expanding? When we say that the universe is expanding, we mean that galaxies that are not bound by gravitation in orbits around each other are rushing away from each other. Our Milky Way galaxy is part of what is called the Local Group of galaxies, containing not only our galaxy and the large spiral galaxy M31 in the constellation Andromeda, but also a number of smaller galaxies, all held to each other by gravitation. The Local Group is not expanding; in fact, the Andromeda galaxy and our own galaxy are moving more or less *toward* each other. But any pair of galaxies that are not bound together like the galaxies of the Local Group are rushing away from each other. This is what is meant by saying that the universe is expanding. It is not that our own galaxy is particularly repulsive; instead, the universe is filled with galaxies rushing away from one another, which as far as we can tell fill all space, with no center and no edge.

We don't really know whether the expansion of the universe will continue forever. It's possible that even galaxies that are not bound to each other in clusters or small local groups will eventually be drawn back together by the gravitational attraction produced by the energy of the whole universe, including the energy locked up (according to Einstein's relation $E = mc^2$) in particle masses. Whether or not this will happen depends on how much energy there is on average in each cubic meter of the universe, on how this energy changes with time, and on the speed at which the universe is expanding. If there is too little energy per volume, and the gravitational field of the universe is too weak to stop the expansion, the universe goes on expanding forever, getting colder and emptier. If there is too much energy in particle masses per volume, then the expansion stops and reverses, recreating the hot dense conditions of the early universe. The choice is between a big chill and a big crunch.

Surprisingly, astronomical observations over the past few years have indicated that the expansion of the universe is not, in fact, slowing down at all, but rather speeding up. If each galaxy had been moving with a constant speed since the beginning, then the distance away from us that any typical galaxy would have reached by now would be proportional to that speed—the faster they go, the farther they get. If, as had been thought, they have been slowing down under the influence of gravitation, then although faster galaxies would be farther from us than slower ones, they would not be as far as they would be if their distance were proportional to their measured speed, because the speeds away from us that we are measuring are the speeds the galaxies had long ago, when the light we observe was emitted by their stars, and if they are slowing down then long ago they would have been moving faster than they have since then.

In fact, two large groups of astronomers (the High-Z Supernova Search Team and the Supernova Cosmology Project) have recently found that distant galaxies are even farther from us than they would be if their distances were proportional to the speeds we observe, indicating that they have been speeding up rather than slowing down since the light we now observe left their stars.[4] About the cause of this acceleration, the best we can say is that in addition to the energy in the masses of ordinary matter, there may be a "vacuum energy" in space itself.

There have been guesses about a vacuum energy (sometimes called "dark energy") since Einstein first turned his attention to cosmology,[5] and it has long been understood that vacuum energy would produce a kind of antigravity, leading to a repulsive force

4. The speeds of distant galaxies are measured by the Doppler effect, the increase of the wavelengths of the light from the stars of the galaxy. It is much harder to measure the distance of the galaxies. What is needed is a "standard candle," some sort of object that can be seen at great distances, and that emits light at a known rate, so that we can judge its distance from its apparent brightness. The discovery that the expansion of the universe is accelerating was made possible by the use of a new standard candle: the thermonuclear explosions of white dwarf stars, explosions known as Type Ia supernovas.

5. Einstein in 1917 introduced a modification in the equations of his 1915 General Theory of Relativity, known as a cosmological constant, which was completely equivalent to attributing to empty space an energy per volume that is the same everywhere and at all times. After the advent of quantum mechanics, it was realized that the uncertainty principle does not allow fields to have any constant value even in supposedly empty space, so that fluctuations in these fields would inevitably also contribute to a vacuum energy. In some modern theories there is an additional vacuum energy that evolves with the universe; this is sometimes known as quintessence.

among the galaxies at great distances from each other, but so far we have had no reliable way of predicting the size of this effect, or how it evolves with time. This is a fundamental problem for physics as well as cosmology. One of the drawbacks of our present string theories is that they either make no prediction about the vacuum energy or predict a value much too large to be consistent with observation. Not having a final theory thus stands in the way of predicting the future of the universe as well as its past.

Because we judge the distances of galaxies by the apparent brightness of things they contain, distant galaxies would also look farther away than they actually are if light from these galaxies has been dimmed by intergalactic dust along our line of sight. Fortunately there is a way of distinguishing this dimming from the effect of a vacuum energy. The energy per volume in particle masses was larger at early times than it is now, when the matter of the universe has become less tightly packed, while the vacuum energy per volume is expected to have been (at least roughly) constant; so during very early times mass energy would have dominated over vacuum energy.

This means that, despite the presence of vacuum energy, the expansion of the universe at very early times would have been slowing down, not speeding up. Thus, although (as has been observed) moderately distant galaxies would be farther away than they would be if their distance were proportional to their observed speed, an extremely distant galaxy, whose speed we measure as it was at very early times, would be closer to us than it would be if its distance were proportional to its speed, because it would have slowed down for a long while before it began to speed up. Light from such an extremely distant galaxy would therefore seem brighter than it would if the galaxy's distance were proportional to its speed, an effect that could not be produced by intergalactic dust. Early this year astronomers announced that just this effect had been found, through the study of a supernova that had been observed in 1997 in a galaxy moving away from us at an exceptionally high speed.

If the acceleration of the universe will continue, as it would if it were caused by a constant vacuum energy, then we are definitely in for a big chill, not a big crunch. Also, under reasonable assumptions about the rate of acceleration, we would be surrounded by what is called an event horizon: any galaxy beyond a certain distance would be forever unobservable. Even if we were to set out now in a spaceship traveling at the speed of light, we could never catch up with galaxies beyond the event horizon, because the longer we travel toward them the faster they would be going away from us. Further, as time passes, more and more galaxies leave our event horizon, and become forever out of reach. If you want to explore any part of the universe outside the Local Group (the group of galaxies held together by gravity that I mentioned earlier, including our galaxy and the Andromeda galaxy) then you will have to do it in the next hundred billion years or so. After that it will be too late, and you will never be able to visit any galaxy beyond the Local Group.

It may be that the entire discussion so far has been parochial. Some of the most interesting ideas in modern cosmology involve the possibility that the expanding "big bang universe" that I have been describing is just one episode in a much larger multiverse in which big bangs go off here and there all the time. It may be that big bangs like our own have happened infinite numbers of times in the past, and will go on happening again and again. We're very far now from knowing whether this is true. The greatest obstacle, I think, is not a lack of suitable astronomical observations, because it's hard to see any way that observational astronomy could settle this issue, but rather a lack of a fundamental physical theory. The ideas about multiple big bangs grew out of speculations about certain fields that might appear in a fundamental physical theory, but so far they are only speculations. When we have discovered what I call the final theory, whether it's a string theory or whatever it is, one of the things we will learn is the answer to the question of whether our big bang accounts for the whole universe. But even if it does not, we will remain forever trapped in our own big bang.

What future is there for us in this ever-expanding universe? In *The Time Machine,* H. G. Wells tells how the Time Traveler journeys forward thirty million years, to stand on a beach under a sun growing cold, with the sea beginning to freeze and the highest form of life a round thing the size of a football, hopping fitfully about on the beach. We know better now. The Sun is what is called a main sequence star, which means that it is getting its energy through nuclear reactions in which hydrogen fuses into helium at the solar core. As the hydrogen gets used up, the Sun will at first heat up, not cool down. Our oceans will boil in about three and a half billion years.[6]

Eventually, the Sun will swell into a red giant, with hydrogen no longer available at the core. If you would like to see what the Sun will look like then, take a look at the constellation Orion. In one corner of the constellation, there's a distinctly reddish star, Betelgeuse. Our Sun will become a red giant like Betelgeuse in about seven billion years.

We don't really know whether the Earth will be destroyed then. It may be that it will experience a drag from the expanding atmosphere of the Sun, like the atmospheric drag that brings down artificial Earth satellites after a few years, and that this drag will cause the Earth to spiral into the Sun. Or it may be that, as the solar wind takes more and more mass from the Sun, the Earth's orbit will expand, and the Earth will escape being drawn into the sun. But with the oceans gone, who will care?

Eventually the Sun will become less luminous, turning into a dwarf star, probably what is called a white dwarf, about as big as the Earth is now. Long before then either our species will have dis-

---

6. The numbers in this article are based on the calculations of many physicists and astronomers, as summarized in two excellent review articles: Freeman J. Dyson, "Time without End: Physics and Biology in an Open Universe," *Reviews of Modern Physics* 51, no. 3 (July 1979): 447–460; and Fred C. Adams and Gregory Laughlin, "A Dying Universe: The Long-term Fate and Evolution of Astrophysical Objects," *Reviews of Modern Physics* 69, no. 2 (April 1997): 337–372.

appeared or we will have colonized other parts of the universe, perhaps taking the Earth with us. Right now it doesn't seem that we're very active in colonizing even the planets in our own solar system. The human race is not living up to the expectations of science fiction authors.

My guess is that, although expeditions will plant scientific stations on Mars, the asteroids, and the moons of Jupiter and the other outer planets, and perhaps eventually on planets around other stars, we're not going to be colonizing any of them for a very long time. In part, my argument for this is based on our experience with Antarctica. There are scientific stations in Antarctica, but does anyone think of forming an economically self-sufficient permanent colony there? Yet, compared to Mars, Antarctica is heaven. So as long as we're not colonizing Antarctica, I can't conceive of any reason why we would colonize Mars or the moons of Jupiter, let alone planets of a distant star. But we will have a motive for colonizing other planets eventually, at least when the oceans start boiling in three and a half billion years. So the future of humanity may eventually depend on the future of more of the universe than just the solar system.

New stars will continue to form and provide sources of nuclear energy for quite a long while, because there's lots of interstellar gas and dust in our galaxy and other galaxies that hasn't yet formed into stars. But star formation will be over in about a trillion years.

At this point, I have to stop using words like billion and trillion, and start to use the common language of science, expressing large numbers as powers of ten. A trillion is 10^{12}, which means that it is ten multiplied by itself twelve times. Or you can think of 10^{12} as meaning a one followed by twelve zeroes. Generally speaking, the lower the mass of a star, the more slowly it evolves. Energy production in the lowest-mass stars will be over in about 10^{14} years—one hundred trillion years. Then the galaxy will contain only brown and white dwarf stars, no longer fueled by nuclear reactions, plus a few neutron stars, which are essentially just large atomic nuclei a few miles in diameter, and black holes.

There will be some rebirth of stars caused by the coalescence of these relics, and so nuclear reactions will now and then start going again. But after a while this too will be pretty well over, because the galaxies will evaporate. Most stars are now held in the gravitational field of their galaxy, and don't have a chance to escape from the galaxy, just as the Moon doesn't have a chance to escape from the gravitational field of the Earth. But every once in a while, two stars come close enough together so that one of them picks up enough speed to reach escape velocity and leave its galaxy. This process is very slow because stars don't come close to each other very often. But the galaxies will have pretty well evaporated in 10^{18} years, ending the rebirth of stars.

In the final chapter of my 1977 book *The First Three Minutes*, I said that the universe "faces a future extinction of endless cold or intolerable heat," and I concluded that "there is not much of comfort in any of this." Freeman Dyson, one of the most perceptive and imaginative scientists I know, was moved by this remark to write an article[7] with more optimistic conclusions. He acknowledged that there was no hope for us in a big crunch. But he argued that in the case of a big chill (which now seems in store for us)— even though in the deepening cold, physical processes would be slower and slower—our descendants could slow down their thoughts even more, so that they would always have an infinite number of thoughts left to think.

Dyson also thought of imaginative slow sources of energy, as for instance cold fusion, i.e., nuclear fusion near or below room temperature. Of course, most scientists are skeptical about the sort of cold fusion that has been in the news in recent years, but Dyson's version of slow fusion really would happen, if given enough time. The atomic nuclei in molecules or crystals are separated by barriers of electric force, which normally keep them from reacting with each other, but through which in fact they can slowly leak, so that coming into contact they can trigger nuclear reactions. In this way

7. Dyson, "Time without End."

a mass of carbon in a burned-out star will eventually turn to iron, releasing nuclear energy. But this is a process so slow it defies the imagination. Dyson estimated that at low temperatures the fusion of carbon into iron would take about 10^{1500} years.

Dyson may have been too optimistic. There are good reasons to think that the nuclear particles that make up most of the mass of all ordinary matter will decay into lighter particles long before cold fusion can occur. The nuclear particles seem stable under normal circumstances, but it is widely expected that they actually decay,[8] with a half-life in the neighborhood of between about 10^{32} years and 10^{37} years, much shorter than the time needed for Dyson's cold fusion.

This is another thing that we can't predict without a more comprehensive physical theory, but there is a chance that if matter does decay in this way, then the decay can be observed. That may seem absurd, but you don't look for this decay by getting one nuclear particle and waiting 10^{32} years. Instead, you get more than 10^{32} particles, which weigh about a hundred tons, and you wait a few years. Well, it's a little harder than that. But with great big tanks of water in underground mines, like one in Japan, there is a good chance that if nuclear particles decay with a rate at the high end of what we expect, it will be discovered in the next few years.

These speculations about the decay of nuclear particles were fairly new when Dyson wrote his article, so he did not give them much attention. But if they are correct, then after about 10^{40} years there will be no atomic nuclei left, and hence no atoms or molecules. The only things left in the universe will be radiation and

8. This is often called proton decay, even though there are neutrons as well as protons inside atomic nuclei, because free neutrons decay quite rapidly into protons, in about ten minutes, while free protons have such long lifetimes that none has ever yet been observed to decay. But neutrons inside ordinary nonradioactive nuclei (like the nuclei of the most common isotopes of most elements) are expected to decay about as slowly as protons, so the discovery of the decay of neutrons inside such nuclei would be just as exciting as the decay of protons.

maybe a few electrons, neutrinos, antielectrons, and antineutrinos. The universe will be a very dull place.

That's not quite true, because the universe will still also contain black holes. It is not that black holes are not made out of nuclear particles. Black holes are made of nuclear particles and electrons, just like ordinary stars, but a black hole is so compact that its gravitational field stops light from escaping from its surface, while any light that escapes from just outside its surface is slowed so much that, to an outside observer, time on the surface of the black hole seems to have stopped. From the point of view of an observer falling into a black hole, nuclear particles will seem to be decaying with the same half-life that we will observe on Earth (or hope we will), but from far outside, they seem to live much, much longer. But the black hole itself radiates away its energy, so that a black hole with the same mass as the Sun will in any case disappear in 10^{66} years or so. A galaxy-sized black hole lives longer, perhaps 10^{100} years. Eventually the black holes, too, will all be gone. If the Time Traveler journeyed that far forward in time, he would find no beach, no planets, no stars, no atoms, nothing but "creeping murmur, and the poring dark."

The view of the future of humanity that I have presented here is not entirely jolly. Putting aside our dismal predictions about the distant future of the universe, we may, in the near term, be able to discover the fundamental laws of nature, but we will never know why they are true. As far as we can tell, these laws will be quite impersonal, not showing any sign of concern for human beings. And although we may learn how we have come to have the values we have, and scientific knowledge will doubtless continue to improve our ability to get the things we value, nothing in science can ever tell us what we ought to value.

This, I suppose, is a rather tragic view of human life. But it did not originate with scientists. It is beautifully expressed, for instance, in Shakespeare's plays. Prospero could almost have been thinking of the decay of protons or black holes when he described how everything would "dissolve, and, like this insubstantial pageant faded, leave not a rack behind." The loves of Titania and Demetrius remind us how much accident there is in determining what it is that we will value. And for most of us, as for Shakespeare, none of this cosmological angst is as important as the fact that for each person, the universe will effectively cease to exist in at most about 10^2 years. As Guiderius sings in *Cymbeline*, "golden lads and girls all must, as chimney-sweepers, come to dust." But our tragedy is not like the tragedy of Lear or Othello. Their tragedy is in Shakespeare's script. Our tragedy is that there is no script.

Or rather, we have to write the script ourselves. We can decide for ourselves which of our inherited values to hold on to, such as loving each other, and which to abandon, like the subordination of women. And there are new values that we can invent. Though aware that there is nothing in the universe that suggests any purpose for humanity, one way that we can find a purpose is to study the universe by the methods of science, without consoling ourselves with fairy tales about its future, or about our own.

This essay was criticized in two letters, published in the *New York Review* in March 2002. One letter, by Ruth Hubbard, Emeritus Professor of Humanities at Harvard, asked if I did not realize that I had just offered another fairy tale, more consoling or less, depending on one's taste. The other letter, by Mark Wagner, Professor of Humanities at Nichols College, criticized my statement that "there is nothing in the universe that suggests any purpose for humanity." I answered by spelling out the degree of certainty or uncertainty that should be attached to different predictions about the fate of the universe, and continued as follows: "All this is neither a fallacy nor a fairy tale—it is just the best we can do right now to predict what will really happen to the universe. I find some consolation in the ability of humans to ponder the future in this way, though Professor Hubbard is quite right to say that this is a matter of taste. But whatever one's taste in consolation, surely it is the nobler path to confront what scientific research reveals about the universe without wishful thinking, neither demanding that it should console us nor that it should celebrate the individual, as Professor Wagner would like, but asking only whether it is likely to be true."

5

Dark Energy

The year 2001 was not only the first of a new century; it was also the 100th anniversary of the Nobel Prize. In December there was a grand reunion of laureates in Stockholm, and that month the *Times Higher Education Supplement* of London joined in the festivities with a special section, *100 Years of the Nobel,* with laureate Harry Kroto as guest editor. Nadine Gordimer, Günter Grass, Leon Lederman, James Mirries, John Polanyi, Joseph Rotblat, and James Vane as well as Kroto all took the opportunity to comment on public issues facing our world, but I decided instead to report on a scientific problem having to do with the whole universe. It is the problem of dark energy, a tiny energy resident in space itself, that I mentioned briefly in the previous essay in this collection.

This problem had been on my mind for some time. In 1988, in preparation for a series of lectures at Harvard, I surveyed various proposed explanations of why the dark energy, if any, was too small to have been detected, and concluded that only one explanation seemed to work, the multiverse idea mentioned in the preceding essay. In 1998, Hugo Martel, Paul Shapiro, and I at the University of Texas made an estimate of the dark energy that would be expected according to this idea; we found that it should be large enough to produce a measurable acceleration of the expansion of the universe. And then in the same year two groups of astronomers found that the expansion of the universe is indeed accelerating.

My topic is the energy of empty space. This may seem like an exceptionally unpromising subject. Isn't empty space just, well, empty? Not quite. In fact, our failure to understand the energy of empty space is the most important roadblock today to further progress in both fundamental physics and cosmology.

Heisenberg's uncertainty principle tells us that space cannot be truly empty. In its most familiar form, this principle states that it is impossible for the position of a particle and the rate of change of its position—the particle's velocity—both to have definite values at the same time. A similar uncertainty principle applies to fields, such as electric, magnetic or gravitational fields. It is impossible for a field and its rate of change at any given point both to have any definite values at the same time. In particular, it is not possible for any of these fields to remain zero even in supposedly empty space; if a field happened to be zero at one instant, then its rate of change could have any possible value, so the field could have any possible value at any later instant. These continually fluctuating fields give an energy to any volume of space itself, proportional to the volume, whether or not the volume contains ordinary matter.

Who cares? Normally, it is not energy that matters, but *differences* in energy. To learn how much energy we can extract from a ball rolling down a hill, we have to ask what is the difference in the gravitational energy of the ball on top of the hill and at its bottom. But the energy in space is there all the time, whatever else is present, and therefore has no effect on the differences in the energies of any two states.

There is just one thing in nature that does respond to energy itself, and not just to differences in energies: gravitation. The gravitational field of the Sun receives contributions not only from the Sun's mass but also from its heat energy, so that to stay in its orbit the Earth must go around the Sun slightly faster than it would if the Sun were cold. There is a lot of space in the universe, and its energy contributes to the cosmic gravitational field, which in turn governs the way that the universe is expanding. This has provided an experimental upper limit on the energy of empty space. Depending on its sign, too much dark energy would have either stopped the expansion before now or speeded it up so much that there would have been no time for galaxies or stars to form.

So how much energy do we expect there to be in a cubic foot of empty space? At first sight, it seems that the quantum fluctuations in the electric, magnetic, and gravitational fields would give a cubic foot of vacuum an infinite energy, but this absurd result applies only if one includes the contributions of fluctuations of any possible wavelength, from one foot down to zero. This is not justified, for we really do not know much about the behavior of fluctuations of very small wavelength. If in this calculation we include only wavelengths larger than the shortest distance that has been probed in elementary particle accelerators, about 10^{-16} centimeters, then the calculation gives a finite energy per cubic foot, but an energy that is larger than allowed by observations of how the universe is expanding—too large, in fact, by a factor of 10^{56}! If the dark energy were this large, then the universe would have evolved much too fast to allow time for the appearance of galaxies or stars or life.

This is a puzzle, but not yet a paradox, because there are other possible contributions to the dark energy. Einstein in 1917 proposed introducing a modification to the field equations of General Relativity, known as the cosmological constant, which would be equivalent to giving empty space a constant energy per volume. So if quantum field fluctuations give empty space some huge positive energy, larger by a factor of 10^{56} than is allowed by observation, we can just assume that Einstein's cosmological constant gives an energy that is equally huge, but negative, so that the two energies cancel. But to avoid a conflict with observation, the cancellation would have to be accurate to 56 decimal places, which seems pretty miraculous.

This puzzle provides a good illustration of the style of modern physics. It is not hard to invent a theory of dark energy that would agree with all the data. We can just follow Einstein and introduce a cosmological constant, adjusting its value so that the net dark energy is as small as needed to avoid conflict with observation. But our aim is not just to develop theories that agree with observation; our theories also have to explain why nature is the way it is. So far we have failed to reach this sort of understanding of the energy of empty space.

This mystery has been on physicists' minds for decades. Most of us assumed that there was some fundamental principle, not yet discovered, that required an exact cancellation of the dark energy. We thought that the largest failure of our most ambitious unified theories of gravitation and other forces was that they had given no rationale for such a cancellation. It was a great surprise, then, to learn from astronomical observations in 1998 that the energy of empty space, though vastly less than might have been expected, is apparently not zero. It seems to equal about twice the energy contained (according to the relation $E = mc^2$) in the masses of the constituents of the universe. (This presents us with another puzzle: the energy in each volume of space itself is usually presumed to be at least roughly constant, while the energy per volume in particle masses naturally decreases as the universe expands, so why should they have similar values at this particular moment in the history of the universe?)

This value of the dark energy was inferred from a study of the expansion of the universe, as indicated by the way that distant galaxies are rushing away from us. If the galaxies we observe have been travelling at constant speed since the beginning, then the distance of any galaxy would be proportional to its speed. In the absence of a dark energy, we would expect the galaxies to be slowing down under the influence of their mutual gravitational attraction, so that the speed they had in the time of emission of the light we see now would have been greater than the speed they have had since the light was emitted, and their distances would therefore be smaller than they would if their speeds were constant. In fact, it seems that the distances are larger than they would be if the speeds were constant, indicating that they are not slowing down but speeding up. This is just the effect that would be expected from a positive dark energy. Incidentally, if the expansion of the universe continues to accelerate, then there are galaxies now whose light will never reach us, and eventually all galaxies beyond our Local Group will be invisible to us.

Whether the dark energy is actually zero or has the value indicated by these recent observations, there is no question that it is incredibly tiny compared with what we would have expected from our estimate of the energy in quantum field fluctuations. Many explanations have been proposed. Perhaps some field automatically adjusts its value to cancel almost all of the dark energy. Perhaps some unknown physical principle dictates that the universe is evolving towards a state with zero dark energy, so the dark energy is small now because the universe is already pretty old. Perhaps our big bang is just one episode in a "multiverse" in which countless big bangs have occurred and will occur, each with different values for fundamental physical constants like the dark energy. In that case, any creatures that ask about the dark energy must be in a big bang where the dark energy happens purely by chance to be small enough to allow time for them to evolve to the point of asking the question. Whatever the true explanation, it is bound to be interesting.

———

Since this essay was published, the study of the cosmic microwave radiation background (a faint whisper of radio waves left over from a time when the big bang was only 380,000 years old), has provided confirmation of the existence of dark energy. The hottest problem right now in astronomy is to measure whether the amount of dark energy in each volume of space remains constant, or evolves with time as the universe expands. In the United States, NASA and the Department of Energy are sponsoring a joint satellite mission to measure the amount of dark energy at various times over the past 10 billion years, and astronomers have plans for making similar measurements using ground-based telescopes, such as the Hobby-Eberly telescope in West Texas.

———

6

How Great Equations Survive

There is one demand that editors regularly impose on scientists who write for the general public: DO NOT USE EQUATIONS. It therefore must have taken some courage for Graham Farmelo, a physicist and writer about physics, to put together a collection of essays, *It Must be Beautiful—Great Equations of Modern Science,* that deal specifically with some of the iconic equations of science. The book was published by Granta Books in 2002, and even exhibits half a dozen or so actual equations. Among the contributors were friends of mine, including the historians Peter Galison and Arthur Miller and the physicists Roger Penrose, Christine Sutton, and Frank Wilczek, and others whom I don't know personally but whose writings I admire, including the population biologists Robert May and John Maynard Smith.

I accepted the task of writing the Afterword, commenting on how great equations survive in our theories of nature, even when some of the motivating ideas of the scientists who introduced them are no longer credited. The essay touches on Maxwell's equations of electricity and magnetism, and Einstein's equations of General Relativity, but gives special attention to Dirac's wave equation of the electron. This was due partly to a personal peeve. When, as a student, I learned about Dirac's equation, the rationale given for the equation was just the one that had motivated Paul Dirac, and I suspect that it is still taught this way, even though this makes no sense in the light of current understandings. But the equation survives anyway, as part of our best modern theories, and one way or another always will. This is the sort of thing I wanted to explain in this essay.

It is terribly difficult for us to put ourselves in the frame of mind of those who lived in past centuries, but many of their artifacts—buildings, roads, works of art—have survived, and some are even still in use. In the same way, though it is often hard for us to understand the thinking of past scientists who did not know what we know, the great equations that bear their names—Maxwell's equations for the electromagnetic field, Einstein's equations for the gravitational field, the Schrödinger equation for the wave function of quantum mechanics, and the other equations discussed in this book—are still with us, and are still useful. These equations are monuments of scientific progress, just as cathedrals are monuments to the spirit of the Middle Ages. Will there ever come a day when we do not teach these great equations to our students?

Although these equations are permanent parts of scientific knowledge, there have been profound changes in our understanding of the contexts in which they are valid, and of the reasons why they are valid in these contexts. We no longer think of Maxwell's equations as a description of tensions within the ether, as Maxwell did, or even as an exact description of electromagnetic fields, as his fellow physicist Oliver Heaviside did. We have known since the 1930s that the equations governing electromagnetic fields contain an infinite number of additional terms, proportional to higher and higher powers of the fields and of the frequency with which the fields oscillate. These additional terms are tiny at the frequencies and intensities we are used to in visible light, but at much higher intensities can lead to an appreciable scattering of intersecting rays of light. Maxwell's theory is an effective field theory, a theory that is a good approximation only for fields that are sufficiently weak and slowly varying.

The additional terms that must be added to Maxwell's equations arise from the interaction of electromagnetic fields with pairs of charged particles and antiparticles that are continually being produced from empty space and then annihilating again. The calculations in the 1930s of the largest of these additional terms were carried out using quantum electrodynamics, the quantum theory of electromagnetism, electrons, and antielectrons. Quantum electrodynamics is itself not the final answer. It arises from the equations of a more fundamental theory, the modern Standard Model of elementary particles, in an approximation in which all energies are taken to be too small to create the quanta of the W and Z fields, the fields that appear in the Standard Model as the siblings of the electromagnetic field. And the Standard Model is not the final answer; we think it is only a low-energy approximation to a more fundamental theory whose equations may not involve electromagnetic fields or W or Z fields at all.

The equations of General Relativity have undergone a similar reinterpretation. In deriving his equations, Einstein was guided by a fundamental insight, the principle of equivalence of gravitation and inertia, which leads to the interpretation of gravitation as a curvature of space-time geometry. But he also introduced an ad hoc assumption of mathematical simplicity, that the equations should be of the type known as second-order differential equations. This means that the equations were assumed by Einstein to involve only rates of change of the fields (first derivatives) and rates of change of rates of change (second derivatives) but not rates of higher order. I don't know any place where Einstein explained the motivation for this assumption. In his 1916 paper on General Relativity he claimed that there was "a minimum of arbitrariness in the choice of these equations," because these were essentially the only possible second-order differential equations for gravitational fields that would be consistent with the principle of equivalence of gravitation and inertia, but at least in that article he made no attempt to explain why the equations had to be of second order. Perhaps he relied on the also unexplained fact that when Newton's theory of gravitation is written in terms of a gravitational field, the equation (Poisson's equation) obeyed by these fields is of second order. Or perhaps he felt that equations of such fundamental importance just have to be as simple as possible.

Today General Relativity is widely (though not universally) regarded as another effective field theory, useful only for distances much larger than about 10^{-33} centimeters, and particle energies much less than an energy equivalent to the rest mass of 10^{19} protons. No one today would (or at least no one should) take seriously any consequence of General Relativity for shorter distances or larger energies.

The more important an equation is, the more we have to be alert to changes in its significance. Nowhere have these changes been more dramatic than for the Dirac equation. Here we have seen not just a change in our view of why an equation is valid and of the conditions under which it is valid but also a radical change in our understanding of what the equation is about.

Paul Dirac set out in 1928 to find a version of the Schrödinger equation of quantum mechanics that would be consistent with the principles of Special Relativity. The Schrödinger equation governs the quantum-mechanical wave function, a numerical quantity that depends on time and on position in space, and whose square at any position and time gives the probability at that time of finding a particle at that position. The Schrödinger equation does not treat space and time symmetrically, as would be required by Special Relativity. Rather, the rate of change of the wave function with time is related to the *second* derivative of the wave function with respect to position (that is, the rate of change with position of the rate of change of the wave function with position). Dirac noted that the relativistic version of the Schrödinger equation for a particle without spin (the Klein-Gordon equation) is not consistent with the conservation of probability, the principle that the total probability of finding the particle somewhere must always be 100 percent.

Dirac was able to construct a relativistic version of the Schrödinger equation consistent with the conservation of probability, known ever since as the Dirac equation, but it described a particle with a spin equal to one-half (in units of Planck's constant), not zero. This was regarded as a great triumph, for the electron was already known to have spin one-half, through the interpretation of atomic spectra a few years earlier. Moreover, by studying the effect of an external electromagnetic field on his equation, Dirac was able to show that the electron is a magnet with just the magnetic strength that the Dutch experimenters Samuel Goudsmit and George Uhlenbeck had inferred from spectroscopic data, and he was able to calculate the "fine structure" of hydrogen, the tiny differences in energy between states that differ only in the orientation of the electron spin relative to the plane of its orbit. When Dirac died in 1984, one of his obituaries credited him with explaining why the electron must have spin one-half.

The trouble with all this is that there is no relativistic quantum theory of the sort for which Dirac was looking. The combination of relativity and quantum mechanics inevitably leads to theories with unlimited numbers of particles. In such theories the true dynamical variables on which the wave function depends are not the position of one particle or several particles, but fields, like the electromagnetic field of Maxwell. Particles are quanta—bundles of the energy and momentum—of these fields. A photon is a quantum of the electromagnetic field, with spin one, and an electron is a quantum of the electron field, with spin one-half.

After all, if Dirac's arguments were correct, the same arguments would have to apply to any sort of elementary particle. Nothing in Dirac's analysis made use of the special properties of electrons that distinguish them from other particles—the fact, for instance, that electrons are the particles found in orbit around the nucleus in all ordinary atoms. But contrary to what Dirac thought, quantum mechanics and relativity do not forbid the existence of elementary particles with spins different from one-half, and such particles are now known to exist. There is not only the massless photon, with spin one, but massive particles of spin one, the W and Z particles, that seem just as elementary as the electron. There is not even anything in relativistic quantum mechanics that forbids the existence of elementary particles of spin zero. Indeed, such particles appear in our present theories of elementary-particle interactions, and much effort is being expended by experimental physicists to find these spinless particles.

Dirac's theory claimed as its greatest triumph the successful prediction of the antiparticle of the electron, the positron, discovered in cosmic rays a few years later. Dirac had observed that his equation had solutions of negative energy. To avoid a collapse of all atomic electrons into the negative energy states, he supposed that these states were almost all full, so that the Pauli exclusion principle (which forbids two electrons from occupying the same state) would preserve the stability of ordinary electrons of positive energy. The occasional unfilled negative energy state would be interpreted as a particle of positive energy and electric charge opposite to that of the electron, that is, as an antielectron. But from the perspective of quantum field theory, there is no reason why a particle of spin one-half must have a distinct antiparticle. Particles of spin one-half that are their own antiparticle appear in some of our theories, though none has been detected yet.[1] Of course, quantum field theory tells us that an electrically charged particle must have a distinct antiparticle, but this is just as true for particles of spin zero or one (which do not obey the Pauli exclusion principle) as for particles of spin one-half, and an elementary particle of spin one with a distinct antiparticle (the W particle) is now well known experimentally.

Confusion on this point continued for many years after Dirac's work, and may even continue today. In the 1950s a new accelerator called the Bevatron was planned at Berkeley that for the first time would have enough energy to produce antiprotons. The objection was raised that everyone knew that the proton had to have an antiparticle, so why design an accelerator to target this particular discovery? One answer that was made at the time was that the proton did not seem to satisfy the Dirac equation, since it has a magnetic field considerably stronger than Dirac's theory would predict, and if it did not satisfy the Dirac equation, then there was no reason to expect it to have a distinct antiparticle. It was still not understood that Dirac's equation has nothing to do with the neces-

1. Added note: There is now good evidence that neutrinos have a small mass, which requires these spin one-half particles to be considered as their own anti-particles.

sity for antiparticles. And of course, the antiproton was found at the Bevatron.

So why did Dirac's equation work so well in predicting the fine structure of the hydrogen atom and the strength of the electron's magnetic field? It happens that the fusion of quantum mechanics with Special Relativity requires that a field whose quanta have spin one-half and interact only with an external electromagnetic field must satisfy an equation that is mathematically identical to the Dirac equation, though it has a quite different interpretation. The field is not a wave function—it is not a numerical quantity, like the Schrödinger wave function, but a quantum mechanical operator that changes the number of particles in physical states on which it acts, and it has no direct interpretation in terms of the probabilities of finding the particle at different positions. By considering the action of this operator on states containing a single electron, one can calculate the particle's magnetic strength and the energies of the states of this particle in atoms. Because the equation for the electron field operator is mathematically the same as Dirac's equation for his wave function, the results of this calculation turn out to be the same as Dirac's.

All this is only approximate. The electron also interacts with the quantum fluctuations in the electromagnetic field, so its magnetic field and its energies in atomic states are not precisely equal to those calculated by Dirac, and it also has nonelectromagnetic weak interactions with the atomic nucleus. But these are small effects in ordinary atoms. Although calculations of atomic structure based on Dirac's equation are only approximate, they are very good approximations, and continue to be useful.

So it goes. When an equation is as successful as Dirac's, it is never simply a mistake. It may not be valid for the reason supposed by its author, it may break down in new contexts, and it may not even mean what its author thought it meant. We must continually be open to reinterpretations of these equations. But the great equations of modern physics are a permanent part of scientific knowledge, which may outlast even the beautiful cathedrals of earlier ages.

7

On Missile Defense

In one way or another, I have been involved with ballistic missile defense for almost half a century. I was a defense consultant working in the early 1960s on the problem of discriminating decoys from warheads carrying real bombs; then I collaborated on a book sponsored by Senator Edward Kennedy opposing the Nixon administration's "Safeguard" missile defense system; later I consulted for a while at the Arms Control and Disarmament Agency assessing Soviet antimissile capabilities; and since then I have written about missile defense on my own in various unclassified publications.

In August 2001 President George W. Bush announced that the United States would soon formally withdraw from the 1972 antiballistic missiel (ABM) treaty with the Russians that had limited development and deployment of ballistic missile defenses, and the Senate Foreign Relations Committee decided to hold hearings on this issue. I was invited by Peter Zimmerman of the Committee staff to testify, but the hearings were cancelled after the disaster of September 11, 2001, and congressional opposition to the President's initiative collapsed. This article is an expanded version of the testimony that I had prepared but never delivered. It was published in the *New York Review of Books* in February 2002, and reprinted in *The Best Science and Nature Writing 2003.*

On December 13, 2001, President Bush announced that in six months the United States would withdraw from the 1972 ABM treaty, a treaty that limits the testing and prohibits the deployment of any national missile defense system by Russia or the United

States. The stated reason for this decision was that the United States needs to develop a system that would protect us from attack by intercontinental ballistic missiles launched by terrorists or by a so-called rogue state. The United States has not yet withdrawn from the treaty; this is the formal six months' advance notice that is required by the treaty, and the President could still decide not to withdraw, but it is hard to imagine that anything could happen before June 2002 that would change his mind.

The arguments by scientists and members of Congress that the United States could continue an active program of developing and testing missile defense systems without abrogating the ABM treaty now seem moot. But the issue of whether to actually develop and deploy a national missile defense system is not moot, and will not be settled even after the treaty is abrogated. Requests for missile defense funding will come up again in Congress in mid-2002, and in subsequent years. We can anticipate a continuing national debate about whether the United States should seek to develop and deploy a national system of defense against intercontinental ballistic missiles.

Few of the arguments in this debate will be new. Indeed, it is hard to remember a time when the United States has not been arguing about a national missile defense program.[1] Almost half a century ago, in the Eisenhower administration, the Army proposed to convert the old Nike antiaircraft system to an antimissile system called Nike Zeus, which would send radar-guided nuclear-armed rockets to intercept Soviet warheads as they plunged through the atmosphere toward U.S. cities. It had obvious failings: the nuclear blasts from the first successful interceptions could put our radars out of action, and the stock of interceptor missiles could be exhausted if the enemy missiles carried several light decoys along with each warhead.

---

1. An excellent and evenhanded account of the Bush administration's missile defense plan as well as earlier missile defense proposals is given by Bradley Graham in *Hit to Kill: The New Battle over Shielding America from Missile Attack* (Public Affairs, 2001).

In the Kennedy administration the Nike Zeus plan was upgraded to a two-tier project called Nike X. Long-range nuclear-armed missiles called Spartans would attempt to intercept Soviet missiles while they were still coasting above the Earth's atmosphere; short-range Sprint missiles would then deal in the atmosphere with those warheads that had survived the Spartan attack. As a member of the JASON group of defense consultants, I worked in the 1960s on the problem of discriminating decoys from warheads, and learned how difficult it is. Like others before me, I gradually also became influenced by a powerful argument against deploying any missile defense system: that in the conditions of the times, it would simply induce the Soviets to increase their offensive intercontinental missile forces, leaving us worse off than before.

Despite such arguments, the Johnson administration came under powerful political pressure to go ahead with some sort of missile defense. In 1967 Defense Secretary Robert McNamara gave a remarkable speech in which he explained all the reasons against deploying a national missile defense, and then concluded that the Johnson administration would go ahead anyway with a limited antimissile system, now to be called Sentinel, which would protect our cities only from attack either by accident or by what was then considered to be a rogue state, China.

To everyone's surprise, the most effective opposition to the Sentinel system did not come from experts who criticized its effectiveness or worried about arms control, but rather from citizens who simply did not want nuclear-armed defensive missiles in their neighborhoods. In response to this opposition, the Nixon administration moved the proposed Sprint missile sites away from cities and renamed the system Safeguard. Its declared purpose was now to defend our offensive missile silos instead of our cities against a missile attack. This was intended to defuse worries about strategic stability—protecting our missile silos instead of cities would not make it necessary for the Soviets to increase their forces in order to maintain their ability to retaliate for a U.S. first strike. And by protecting our own offensive missiles, Safeguard would reduce any incentive that we might have to launch missiles in a crisis. As explained by Defense Secretary Melvin Laird, "The original Sentinel plan could be misinterpreted as . . . and in fact could have been . . . a first step for the protection of our cities."[2] But in fact there was little technical difference between the Sentinel and Safeguard systems, except that Safeguard would have less effect on suburban real estate values.[3]

The Safeguard system was scotched by doubts about its effectiveness (especially concerning the vulnerability of its radars) and fears about its cost. In 1972 the Nixon administration and the Soviet Union signed the ABM arms control treaty. It limited defenses against ballistic missiles to one hundred interceptors at each of two sites, later reduced by mutual agreement to one hundred interceptors at one site. The site could be located to protect either the national capital or a field of offensive missiles. This would allow the Soviets to maintain their rather primitive Galosh missile defense system around Moscow, while the United States could proceed with the declared aim of the Safeguard system and defend the intercontinental ballistic missile field in North Dakota.

To guard against surprises, the treaty also contained a clause that banned developing, testing, or deploying "ABM systems or components which are sea-based, air-based, space-based, or mobile land-based,"[4] a clause that later came under special attack by proponents of missile defense. Despite the proclaimed need for defense of our offensive missiles, neither the Nixon administration nor any following administration maintained the ABM de-

---

2. Statement before the Senate Armed Services Committee, March 19, 1969.

3. For contemporary arguments against deploying the Safeguard system (including an article of mine), see *ABM: An Evaluation of the Decision to Deploy an Antiballistic Missile System,* edited by Abram Chayes and Jerome B. Wiesner (Harper and Row, 1969).

4. The texts of various arms control treaties can be found in *Nuclear Arms Control: Background and Issues,* prepared by the Committee on International Security and Arms Control of the National Academy of Sciences (National Academy Press, 1985).

fense of the North Dakota missile field that was allowed under the treaty.

There matters remained until the Reagan administration. It is said that President Reagan was converted to missile defense on a visit to the continental defense headquarters at Cheyenne Mountain, when he was surprised to learn that the United States had no ability to shoot down enemy missiles attacking our country. Be that as it may, in 1983 he announced plans for a Strategic Defense Initiative, intended to make nuclear weapons "impotent and obsolete."[5] No longer would the system be limited to ground-based interceptor missiles; there were plans for more adventurous technologies, including satellites carrying X-ray lasers that could burn through the skin of an offensive missile booster in the first few minutes after it was launched. The imagined system soon came to be called Star Wars.

Eventually it became clear even to the enthusiasts in the Reagan administration that the X-ray lasers and other components needed by the Strategic Defense Initiative were beyond current technological capacities. The administration of George Bush Sr. replaced the Strategic Defense Initiative with the Global Protection Against Limited Strikes system, including about one thousand "brilliant pebbles," small space-based interceptor missiles, along with more conventional land- or sea-based missiles. This strategy also led nowhere, and was allowed to lapse in the Clinton administration.

Research and development continued at a more leisurely pace. In 1996 the Department of Defense announced a plan to continue further development of a scaled-down missile defense system for three years, after which a decision would be made whether or not to deploy the system within the following three years. The National Missile Defense System under study was now limited to a single kind of interceptor missile. Instead of a nuclear weapon, it

5. On the Reagan Strategic Defense Initiative, see Frances Fitzgerald, *Way Out There in the Blue: Reagan, Star Wars and the End of the Cold War* (Simon and Schuster, 2000).

would carry an "exoatmospheric kill vehicle" (EKV) weighing about 120 pounds, which would destroy the enemy warhead above the Earth's atmosphere by a direct hit rather than a nuclear blast. If it worked, it would truly be a bullet hitting a bullet.

Then, on August 31, 1998, North Korea surprised the world by launching a three-stage rocket that carried its third stage over one thousand miles before it broke up into pieces and fell into the Pacific Ocean. The missile did not fly far enough to reach any part of the United States, and it could not have carried a nuclear warhead, but its launch put tremendous political pressure on the Clinton administration to do something soon about missile defense.

In July 1999 President Clinton signed a National Missile Defense Act that had been passed by Congress a few months earlier. Like the Johnson administration's Sentinel initiative, this was more of a defense against Republicans than against external threats. The act committed the United States to deploy a national missile defense "as soon as technologically possible." Later that summer the administration settled on the defense system's initial ("C-1") configuration, which remains as a central element of the missile defense system under study by the Bush administration. Twenty (later increased to one hundred) interceptor missiles carrying EKVs would be based at Fort Greely, Alaska, to be guided to their targets initially by five early-warning radars in Alaska, California, Massachusetts, Greenland, and England. Then, later in their flight, they would be guided by a high-frequency battle management radar on Shemya Island in the Aleutians and finally, in the last six hundred miles of flight, by infrared telescopes carried by the kill vehicles.

This geographical deployment was clearly aimed at defense from North Korean missiles. To better protect the east coast of the United States from missiles launched from the Middle East, it would be necessary later to add interceptor missiles at a second site, perhaps in North Dakota or Maine, and also to add additional battle-management radars. The decision to deploy the C-1 system was to have been delayed until 2000, after some of the components of the system had been tested.

The first test of the EKV was made on October 2, 1999. A dummy warhead that had been sent into space by a Minuteman intercontinental ballistic missile launched from Vandenberg Air Force Base in California was hit over the Pacific by an EKV from an interceptor missile fired from Kwajalein Atoll in the Marshall Islands. But this test was much less significant than it may have seemed. The EKV was at first off course, so that its telescopes did not pick up the Minuteman warhead. When the EKV widened its field of view, at first it saw a large bright balloon decoy, and corrected its course, after which it saw the warhead and managed to hit it. If the warhead had not been accompanied by a decoy it might have escaped detection, and if the decoy had looked more like a warhead the EKV would have hit the decoy instead of the warhead. Even so, it seemed that under the right conditions a bullet could hit a bullet.

Then in January 2000 the EKV failed a second test. The krypton gas needed to cool the EKV's infrared telescopes had been blocked by ice in the plumbing, so that the EKV never saw the warhead, and missed it by over two hundred feet. A third test in July also failed, when the EKV failed to separate from its booster. Further, even if all these tests had been successful, three tests were not nearly adequate to test the system.[6] In August 2000 President Clinton finally decided that the Department of Defense should not start preparing the Alaska site for the battle management radar, and he announced that he would leave the decision whether to deploy the missile defense system to the next administration.

President Bush has taken the movement toward national missile defense in a new direction. Where the Clinton plan called for spending $5.75 billion in 2002 for all work on ballistic missile defense, the Bush plan calls for spending $8.3 billion on the same

---

6. A mordant analysis of the limitations of these tests is presented in "Operational Test and Evaluation Report in Support of the National Missile Defense Deployment Readiness Review," by Philip Coyle, then the director of Operational Test and Evaluation of the Department of Defense (August 10, 2000). This report is now available at a Center for Defense Information website, www.cdi.org/news/missile-defense/coyle.pdf.

tasks. The Bush administration assumed that an antimissile system much like the Clinton administration's National Missile Defense C-1 system would be tested not only by engagements between rockets fired from Vandenberg and Kwajelein but also by interceptors fired from Alaska sites, which could later be converted to operational missile defense sites. Also, the Bush administration proposed to supplement this land-based midcourse interception system with an ill-defined mixture of other systems, including possible airborne or spaceborne lasers that would attack enemy missiles during the initial boost phase of their flight.

The Alaska interceptor test site might violate the 1972 ABM treaty; the development and testing of airborne or space-based missile defense systems surely would, long before any actual deployment. But where President Clinton had ruled out a unilateral abrogation of the treaty, President Bush has from the first been eager to free the United States from its restrictions. In August 2001 he said that the United States will withdraw from the treaty at a "time convenient to America."[7] Since then the disaster of September 11 has brought presidents Bush and Putin into closer collaboration, but the Russians have refused to agree to major changes in the ABM treaty, and now the president has given notice of his intention to withdraw from it.

So here we are again, arguing the pros and cons of missile defense. The debate raises three main issues:

Would a missile defense system actually protect the United States against even the sort of attack that might be launched by rogue states like North Korea or Iraq?

It seems to me likely that the problems that bedeviled the early tests of the EKV can all be solved. The fourth and fifth tests in July and December 2001 were successful, though the interceptor booster

7. See David E. Sanger, "Bush Flatly States US Will Pull Out of Missile Treaty," *New York Times*, August 24, 2001, A6.

failed in a test later in December. The big problem, as it has been since the days of Nike X, is that any number of interceptor missiles could be used up in attacking decoys that had been sent by the attacker along with its warheads.

This is a particularly acute problem for such missile defense systems as that planned as the first phase of the Clinton-Bush National Missile Defense, which rely on intercepting warheads in midcourse, above the Earth's atmosphere. Balloons that are deployed in space at the same time as warheads will follow the same trajectory as the warheads until they reenter the earth's atmosphere. They can easily be shaped to look much like warheads to ordinary telescopes, and heated to look like warheads to infrared sensors. It is also possible and probably even easier to make the warhead look like a decoy by putting it in a decoy balloon, or make it invisible by hiding it in a cooled shroud. There has been no realistic test of the ability of an EKV to hit a warhead that is accompanied by such penetration aids.

Much of the technical argument over several decades about the effectiveness of antimissile systems has focused on the question of whether the United States can solve this problem.[8] It may be that at great cost we could develop a midcourse interception system that could defeat any one particular group of penetration aids and warheads, but of course we are not likely to know what aids are chosen by the attacker, and I don't see how we could ever have confidence in our ability to deal with an unknown threat. The attacker always has the last move.

Another way of defeating a U.S. midcourse interception system was mentioned in the report of a "blue ribbon" panel on missile defense convened by Congress in 1998, which was headed by

8. This question is discussed in detail in the Union of Concerned Science/MIT report "Countermeasures: A Technical Evaluation of the Operational Effectiveness of the Planned US National Missile Defense System" (April 2000). This report is now available at www.ucsusa.org/nuclear_weapons_and_global_security/ missile_defense/technical_issues/countermeasures-a-technical.html.

Donald Rumsfeld, and has been emphasized several times since then by Richard Garwin, one of the panel members.[9] Instead of using a rocket that would launch at most a few nuclear warheads, an attacker could use the same sort of rocket to launch hundreds of "bomblets," containing biological warfare agents, such as anthrax spores. Once deployed, the bomblets would be immune to any sort of missile defense now contemplated. This sort of missile could kill even more people than one carrying a nuclear weapon.

None of these objections applies to a missile defense that can damage an attacker's missile while it is still in "boost phase," i.e., during the brief period when it is being accelerated upward, before it has time to deploy warheads, decoys, or bomblets.[10] But the boost phase lasts only a few minutes. A missile attempting to intercept another missile during the boost phase would have to be launched within about six hundred miles of the intercontinental ballistic missile launch site; so this sort of missile defense system would have to be targeted only at one or at most a few particular potential attackers. For example, a sea-based system that would target missiles in North Korea would not protect against missiles launched from China or Russia. Likewise, an airborne laser would not be effective against missiles that at the end of boost phase are still beyond the horizon, which for a missile at an altitude of 120 miles is about one thousand miles away. (The actual range of the laser would be substantially less than this.) An air- or sea-based boost phase intercept system could be vulnerable to preemptive attack (as also would the radars of the Clinton-Bush National Missile Defense system), but unless this attack were very carefully timed, it would trigger a counterattack that would destroy the enemy's offensive missiles while they were on the ground. If it could be made to work, a system of space-based lasers (or "brilliant

9. For instance, see Richard Garwin's Op-Ed article in *The New York Times*, December 30, 2000.

10. A realistic boost phase intercept system is described by Richard Garwin in *Arms Control Today*, September 2000.

pebbles") might be able to provide protection from threats coming from a much larger area, but this technology does not yet exist, and in any case no specific space-based laser system has been proposed by any administration.[11]

Is it plausible that the United States would be attacked by intercontinental ballistic missiles launched by terrorists or a rogue state, or by accident?

The attack of September 11 made it clear (though it was pretty clear before) that there are people in the world who want to damage us. This seems to have shifted public opinion in favor of missile defense, and it stopped moves in the Senate to deny funding for missile defense tests that would violate the ABM treaty. But the attack also demonstrated that there are ways to hurt the United States that do not involve the launch of intercontinental ballistic missiles. Even nuclear weapons could be delivered in many ways, for instance by using trucks or freighters or (as suggested by the Rumsfeld panel) ship-launched short-range missiles. But the intercontinental ballistic missile is not just one among the many vehicles that might be used by terrorists or a rogue state to attack us with nuclear weapons—*it is the least likely vehicle.* Though some terrorists are willing to commit suicide in their attacks, the heads of the nations that harbor them never have been. The leaders of the Taliban did not publicly acknowledge that the September 11 attacks were organized in Afghanistan, and Qaddafi has never admitted that the explosion of a Pan American airliner over Lockerbie was planned in Libya.

11. For a detailed analysis of the prospects for using lasers or other directed energy weapons for missile defense, see "Report to the American Physical Society," by the Study Group on Science and Technology of Directed Energy Weapons, *Reviews of Modern Physics* 59, no. 3, pt. 2 (July 1987). There have been no technological developments since 1987 that would make the pessimistic conclusions of this report out of date.

But unlike such terrorist attacks, an attack by intercontinental ballistic missiles carries an indelible return address. Every launch of such missiles is inevitably detected and its source identified by the fleet of American Defense Support Program satellites. Even granting that a state like North Korea or Iraq might eventually be able to deploy nuclear-armed intercontinental missiles, why would any head of government, however much he may hate us, attack us with intercontinental ballistic missiles, or allow terrorists on his soil to launch such an attack, when he and they could use many other means to deliver nuclear weapons anonymously?

On the other hand, there are circumstances in which the very visibility of intercontinental ballistic missiles might be an advantage. For instance, the United States might not be deterred from its recent actions in Afghanistan or from trying to overthrow Saddam Hussein by a mere threat of nuclear terrorism; but would it risk trying to overthrow a state that had nuclear-armed intercontinental ballistic missiles?

This is a real problem for America, but it is not clear that anything but a perfect antimissile defense would make much difference. If the United States had an antimissile system that had never been used in action, would this give us sufficient confidence to attack a regime that possessed intercontinental nuclear-armed ballistic weapons, especially when we did not know what sorts of decoys their missiles carried?

But it need not come to this. There is another way that the United States can avoid being subject to nuclear blackmail by states like Iraq or North Korea. It is occasionally mentioned in discussions of missile defense, though briefly and perhaps with some embarrassment. It is preemption. (Or, as it is sometimes called, pre–boost phase interception.) If a country like Iraq or North Korea were suspected of having nuclear weapons, and we saw that it had tested a ballistic missile of intercontinental range, would we really watch them begin to erect these missiles without taking steps to destroy them on the ground? These steps need not involve our use of nuclear weapons; cruise missiles are now sufficiently accurate to do the job with conventional explosives. I very much doubt if intercontinental ballistic missiles could be put in place by any state without the United States knowing it, and indeed they would be of no use for nuclear blackmail unless they were known to us.

This leaves a mistaken launch by Russia or China as the only plausible way that intercontinental ballistic missiles might threaten the United States. Here "mistake" might mean anything from a purely mechanical malfunction in a single rocket, to an unauthorized launch by a few madmen of all of the missiles in a submarine or a land-based missile field, all the way up to the launch of a whole arsenal of missiles ordered by a national leader who is under the mistaken impression that his country is under attack.

Launch by mistake is a serious danger, and although it was not mentioned by President Bush at the time he announced his intention to withdraw from the ABM treaty, it had frequently been cited as one reason for building a national missile defense system. Indeed, a large-scale mistaken attack by Russia is the only plausible threat that could not only damage our country but destroy it beyond our ability to recover. Such an attack would be far more devastating than anything terrorists could manage. Russia has some 3,900 strategic nuclear warheads, of which over one thousand are on land- or submarine-based intercontinental ballistic missiles that are ready to be launched at a moment's notice.[12] These missiles are increasingly vulnerable to an American first strike, as the Russian early-warning capabilities get progressively weaker. At least twice, the Russians have mistakenly thought that they were under missile attack: once in March 1983 because a Soviet satellite mistook bright reflections for the launch of five missiles, and again in January 1995 because a U.S. research rocket that had been detected by a Russian radar was interpreted as an incoming American missile. In both cases the Russian launch process

12. For a description of Russian missile forces, see *Russian Strategic Nuclear Forces*, ed. by P. Podvig (MIT Press, 2001).

came within minutes of the point where they would make a decision whether or not to launch a nuclear retaliation.

The danger of a launch during a crisis is made worse by the fact that most Russian and American warheads are MIRVs—multiple independently targeted reentry vehicles. For example, each Russian SS-18 missile carries ten warheads, each of which can be directed to a separate target. Without MIRVs, and with equal numbers of missiles on each side, there would be a disadvantage in striking first. Even if every missile had a 90 percent chance of destroying its target, the side that struck first, even with all its missile forces targeted at its adversary's missiles, would leave its own arsenal empty while its adversary would still have 10 percent of its forces left. But with, say, ten warheads on each missile, the side that struck first with just 10 percent of its forces could destroy 90 percent of the adversary's forces and still have 90 percent of its own forces left.[13] Of course, this reasoning is insane. No one today thinks that either Russia or the United States would plan such an attack. But in some future crisis, with different Russian leaders and with misleading data coming in from early-warning radars—who knows?

The sort of missile defense planned by the Bush administration would not protect us against a massive attack by mistake. Indeed, it has been specifically advertised *not* to be able to defend us from a large-scale Russian attack. (It might not even protect us against a mistaken launch of a few missiles, since Russian missiles are presumably accompanied by sophisticated decoys or other penetration aids; they surely would be if we were to deploy a missile defense system.) With or without the Bush missile defense plan, we have to face the danger of annihilation by Russian nuclear-armed missiles. With the degradation of Russian early-warning capacities

---

13. To be precise, this calculation only applies to immobile missiles that can be targeted in a missile attack. But Russian submarines spend almost all their time tied up at docks, and Russian mobile land-based missiles are generally kept in fixed garrisons, so for these purposes their missiles are immobile.

and the general loosening of Russian society, this danger may be even greater than it was during the cold war. Which brings me to the third and most important issue.

Would a national missile defense system of the sort proposed by the Bush administration help or hurt our national security?

At first sight, this question seems to answer itself. Isn't any missile defense, however ineffective, better than no missile defense at all?

One trouble with this reasoning is that we do not face a fixed threat, a threat independent of what we do about missile defense. True, we are not now in the position we were in during the 1960s and 1970s, when we could reasonably expect that any U.S. missile defense system would be countered by an increase in Soviet offensive missile forces. The current Russian economy would not support an increase in Russia's missile forces and, indeed, the Russians have been eager to reduce their forces, reportedly down to some two thousand or so strategic nuclear warheads. But this is still a force that could destroy the United States, and much else in the world besides. The large size of their arsenal also increases the danger that Russian nuclear weapons or even long-range missiles might be stolen or sold to terrorists or rogue states. I am told that Russia now maintains tight control over its strategic nuclear weapons, but this wasn't true in the early 1990s, and it may not be true in future. There is nothing more important to American security than to get nuclear forces on both sides down at least to hundreds or even dozens rather than thousands of warheads, and especially to get rid of MIRVs, but this is not going to happen if the United States is committed to a national missile defense.

The Russian nuclear force is the sole remnant of its status as a superpower. Whatever good feelings may exist now between us and Russia, any U.S. system that might defend our country against even a few Russian intercontinental ballistic missiles therefore sets a limit below which the Russians will not go in reducing their strategic nuclear forces. Even if Russia is forced by economic pressures to continue reducing its missile forces, it can cheaply maintain its deterrent, although in ways that are dangerous. It could, for example, remove whatever inhibitions it may now have from launching its missiles on a moment's notice. Nor is Russia likely to eliminate its MIRVs if the United States goes ahead with missile defense. The START II treaty was to have eliminated all land-based MIRVs on both sides, but the Russians have already indicated that they will not go through with implementing this treaty if the United States abrogates the ABM treaty.

As for China, it has right now about twenty nuclear-armed intercontinental ballistic missiles, enough for a significant deterrent against any attack from Russia or the United States. A recent National Intelligence Estimate that was leaked to the *New York Times* and the *Washington Post* predicted that if the United States develops a national missile defense, then the Chinese will increase their forces from about twenty to about two hundred missiles.[14] And if China makes this sort of increase in its missile force, then what will Japan and India do? And then what will Pakistan do?

It may seem contradictory to argue that the proposed national missile defense system would probably be ineffective against even a small attack by a rogue state, while also arguing that it would prevent needed reductions in Russian missile forces and promote increases in Chinese missile forces. But each country "prudently" tends to overestimate the effectiveness of any other country's defenses, especially as they may develop in future. The Soviet deployment of a primitive antimissile defense of Moscow was a major factor in America's decision to multiply its warheads by deploying MIRVs, so that Moscow was in more danger after it was defended than before. I remember how in the early 1970s U.S. defense planners became terrified that Soviet antiaircraft missiles might be given a role in defense against intercontinental ballistic missiles,

14. Roberto Suro, "Study Sees Possible China Nuclear Buildings," *Washington Post*, August 10, 2000, A2; Steven Lee Myers, "Study Said to Find US Missile Shield Might Incite China," *New York Times*, August 10, 2000, A1.

something that never happened. Imagine then how Russians and the Chinese defense planners will take account of the unilateral American withdrawal from the 1972 ABM treaty.

If it were possible tomorrow to switch on a missile defense system that would make the United States invulnerable to any missile attack, then I and most other opponents of missile defense would be all for it. But that is not the choice we face. What is at issue is a missile defense system that will take almost a decade to deploy in its initial phase, and then many more years to upgrade to the point where at best it would have some effectiveness against some plausible threats. During all of this time, however, American security will be damaged by measures taken by Russia and China to preserve or enlarge their strategic capability in response to our missile defense.

There is one sort of missile defense that would not raise these problems. As already mentioned, a defense that targets intercontinental ballistic missiles during the boost phase with either missiles or airborne laser beams could only defend against missile launches within a limited geographical area; and it would also be immune to decoys and other penetration aids. We could defend against the launch of North Korean missiles by using short-range missiles based on ships in the Sea of Japan, though to defend against a launch from Iraq or Iran would require cooperation from Turkey or some republic of the former Soviet Union, respectively. Such a defense would have no effectiveness against missiles launched from sites in Russia or China, which during the boost phase would be beyond the range of any missiles or airborne lasers we might deploy. For this reason, although this sort of defense would violate the 1972 ABM treaty, President Putin has already indicated that he would consider revising the treaty to allow it. But a boost phase intercept system would have all the destabilizing effects of other missile defense systems if it were based on satellites, or if it were combined with exoatmospheric midcourse interceptors like those of the Clinton-Bush National Missile Defense proposal.

Developing a national missile defense system would also harm our foreign relations. It would add to the general perception that the United States is unwilling to be bound by international agreements, such as comprehensive test ban treaties or environmental agreements. It would weaken Putin's hand in dealing with Russian ultranationalists. By trying to defend the United States from missile attacks while leaving our allies defenseless, missile defense would tend to undermine alliances like NATO. A boost-phase intercept system would not really be an exception; it is true that the interception of a long-range missile in the boost phase does not depend much on the destination of the missile, but by interrupting the boost it would probably only cause the warhead to fall short, perhaps on an ally, such as Canada or Germany.

A missile defense system would hurt our security in another important way, by taking money away from other forms of defense. We are simply unable to do everything we can imagine that might defend us. We need to upgrade our hospitals to deal with biological attack; improve security along our border with Canada and in our ports; upgrade the FBI computer system; and so on. All of our activities along these lines are constrained by a lack of funds. Legislation to increase funding for homeland defense was blocked in the House of Representatives because the amounts of money requested exceeded the administration's guidelines.

If we are particularly worried (as we should be worried) about terrorist nuclear attacks on the United States, then we ought to give a very high priority to working with Russia and other countries to get rid of the large stocks of weapons-usable uranium and plutonium that are produced by their power reactors. Russia now holds about 150 tons of plutonium and one thousand tons of highly enriched uranium. This material could be used not only to make nuclear bombs, which can be delivered to the United States in all sorts of ways; even a technically unsophisticated terrorist could instead use it to make so-called dirty bombs, in which an ordinary high explosive is surrounded with highly radioactive material that when dispersed in an explosion would make large urban areas uninhabitable.

Unfortunately, this material is not under tight control.[15] Since 1991 there has been a bipartisan Nunn-Lugar Cooperative Threat Reduction Program that among other things aims at improving the security of Russian control over fissionable materials and making Russian plutonium and uranium unusable as nuclear explosives, but this too is not being adequately funded. A bipartisan panel headed by Howard Baker and Lloyd Cutler has called for spending at an average level over the next decade of about $3 billion a year for securing, monitoring, and reducing Russian nuclear weapons, materials, and expertise.[16] The amount in the 2002 budget for these activities is only about $750 million, even after substantial increases by Congress. Comparison of these figures with the $60 billion quoted cost (certain to be greatly exceeded) of a minimum missile defense system gives a powerful impression that the Bush administration and some in Congress are not entirely serious about national security.

I was at a press conference in Washington in November 2001, when the Federation of American Scientists released a letter signed by fifty-one Nobel laureates that opposed spending on national missile defense programs that would violate the 1972 ABM treaty. One of the reporters present asked me why, if the arguments against national missile defense are so cogent, many people in and out of government are for it? It was a good question, and one to which I am not sure I know the answer. There are the usual pressures for

15. For instance, see Steven Erlanger, "Lax Nuclear Security in Russia Is Cited as Way for bin Laden to Get Arms," *New York Times,* November 12, 2001, B1.

16. "A Report Card on the Department of Energy's Nonproliferation Programs with Russia" (U.S. Department of Energy, January 10, 2000), app. A. (The chart from which this number is taken warns that "it is not intended to be of budget quality, nor to imply that the US should be the sole provider of funds for such a program," but in fact contributions to these activities from other countries have been relatively small.) This report is now available at www.google.com/search?q=cache:http://www.seab.energy.gov/publications/rusrpt.pdf.

large military programs that come from defense contractors and from politicians trading on patriotism. The arguments for national missile defense may seem simpler and more straightforward than the arguments against it. But I think there is also a peculiar fascination with anything that projects American power into space. How else explain the idiocy of the International Space Station, or the card tables that were set up at airports during the Reagan administration by people advocating a "high frontier" missile defense program? I have to admit that thoughtlessness is not a monopoly of missile defense advocates. Some opponents of missile defense are automatically against any large military program. In assessing missile defense, or anything else for that matter, there is no substitute for actually thinking through the issues.

In my own field of physics, we make a distinction between applied physics, which is motivated by some social need, and pure physics, the search for knowledge for its own sake. Both kinds of physics are valuable, but not everything pure is desirable. In seeking to deploy a national missile defense aimed at an implausible threat, a defense that would have dubious effectiveness against even that threat, and that on balance would harm our security more than it helps it, the Bush administration seems to be pursuing a pure rather than applied missile defense—a missile defense that is undertaken for its own sake, rather than for any application it may have in protecting our country.

I sent a draft of this article to a friend, a conservative newspaper columnist, who as I knew favored the Bush missile defense program. We had a long talk about it in his office in Washington, and he told me that he thought that my article represented an artificial way of looking at national defense. He summarized this criticism by asking, if you knew that North Korean missiles carrying nuclear weapons had been launched toward the United States, wouldn't you be glad to have some sort of missile defense, even one that might not be completely effective? I didn't have a good answer ready. A few days later, it occurred to me that I might have asked him, if someone were about to hit you over your head with a hammer, wouldn't you be glad that you always wear a steel helmet? Except that you don't. Why is that? But on reflection, that would only have been a way of discounting the danger of an attack by nuclear-armed missiles—it would not have made the more important points that trying to defend against such an attack would be ineffective, and would increase the danger from the nuclear forces arrayed against us.

A few months after this article was published, I heard from Lloyd Doggett, the congressman who then represented the district where I live in Texas. He had quoted from the article on the floor of the House. It was gratifying—my writing doesn't get quoted too often in Congress. He also quoted an incisive statement about missile defense by former Secretary of Defense William Perry: "[A] relatively small deployment of defensive systems could trigger a considerable nuclear arms race." Just so.

As expected, the United States did withdraw from the 1972 ABM treaty in mid-2002. In the following years, the Bush administration continued its deployment of an antimissile system in Alaska, but it never subjected this system to a single realistic test, against a reentry vehicle accompanied with decoys and other penetration aids. It took steps to deploy an antimissile system in eastern Europe as a purported protection against future Iranian missiles, and this soured U.S. relations with Russia, which sees any antimissile defenses, however ineffective, as the beginning of a threat to its own nuclear deterrent. The new president, Barack Obama, has expressed some reservations about the effectiveness of missile defense systems, but it is too soon to tell whether or not he will continue a program of missile defense. I wish that the arguments in this article might have become irrelevant in the years since it was written, but unfortunately they have not.

8

The Growing Nuclear Danger

All my adult life I have been terrified by the danger of nuclear war. There are plenty of other things to worry about—global warming, terrorism, financial meltdowns—but only nuclear war between Russia and America can wreck our countries, and perhaps all civilization, beyond their ability to recover. At the time I am writing this, news commentators are wondering why we did not realize how unstable the world financial system had become. I fear that at some time in the future, when cities on both sides are in ruins, we may be wondering why we did not realize how unstable was the nuclear stand-off between Russia and America.

This danger never seems to figure in political campaigns, but I can't get it out of my mind. I don't see how Russia and America can abandon their nuclear weapons altogether, as that would put us at the mercy of any third country with a few nuclear bombs, but surely we can and should drastically reduce the enormous nuclear arsenals on both sides. We also ought not to undertake new nuclear weapons programs, which would stir up other countries' interest in and excuses for developing new nuclear weapons of their own.

For these reasons, I was not a little troubled when I learned of the plans of the Bush administration, set out in its *Nuclear Posture Review* of January 9, 2002. Although there would be a reduction in the number of operationally deployed nuclear weapons, the thousands of retired warheads would not be destroyed, and could easily be put back into service. The *Nuclear Posture Review* also called for the development of new nuclear warheads, designed to attack underground facilities.

A short while after news of the *Nuclear Posture Review* was leaked to the press, I received a call from Peter Zimmerman, a member of the staff of the Senate Foreign Relations Committee, asking me to testify on this Review before the Committee later that Spring. As already mentioned in connection with the foregoing article in this collection, on missile defense, I had earlier been recruited by Zimmerman to testify on missile defense before the same committee, but those hearings had been cancelled after the attack of September 11, 2001. Missile defense is something I have worked on and thought about over the years, but some of the issues now raised by the *Nuclear Posture Review* were new to me, so I tried to get the Committee to put someone else in my place, without success. Fortunately I was able to get invaluable information and insights from friends associated with the Federation of American Scientists and the Union of Concerned Scientists. The *Nuclear Posture Review* is classified, and though I had an active security clearance, I wanted to be able to testify about it freely in open hearings, so I did not ask to see the full classified version of the posture review. However, there were available an unclassified briefing on the Review by Assistant Defense Secretary Crouch and an unclassified transmittal letter by Defense Secretary Rumsfeld, as well as excerpts from the Review that had appeared in the press, so I thought I knew enough about the Review to criticize it.

The hearings were held on May 16, 2002. Presiding were Senators Joseph Biden, chairman of the Committee, and Richard Lugar, ranking minority member. Both senators impressed me with their seriousness in dealing with an issue that probably did not much interest their constituents. I knew one of the other witnesses, Dr. John S. Foster, Jr., from my days as an active member of the JASON group of defense consultants. Foster had been Director of the Lawrence Livermore Laboratory and U.S. Director of Defense Research and Engineering, and had served on countless advisory boards. I had always regarded him as a hardliner on defense issues, and was not surprised to hear that he generally supported the Bush administration's nuclear policies. Another witness who took a similar line was Admiral William A. Owens, former Vice Chairman of the Joint Chiefs of Staff. There were also two witnesses whose views were closer to my own: Joseph Cirincione, Director for Nonproliferation at the Carnegie Endowment for International Peace, and Loren Thompson, Chief Operating Officer of the Lexington Institute. I had dinner with Cirincione the night before the hearings, and took the opportunity to prepare for my testimony by picking his brains throughout the meal. The following article is largely based on my testimony at these hearings, expanded to include comments on the arms control treaty signed by Presidents Bush and Putin a week later. It was published in the *New York Review of Books* in July 2002.

One point in this article mirrors something that happened at the hearing. Either Dr. Foster or Admiral Owens remarked that we had to keep a large nuclear arsenal because the future was uncertain. As mentioned in this article, President Bush made a similar argument after signing the treaty with Putin. The answer to this given in my article is the same one I gave at the hearing: Yes, the future is uncertain, but in many of the uncertain future contingencies, the enormous Russian and American nuclear arsenals would lead the world to disaster.

The United States possesses an enormous nuclear arsenal, left over from the days of the cold war. We have about six thousand operationally deployed nuclear weapons,[1] of which roughly two thousand are on intercontinental ballistic missiles, thirty-five hundred on submarine-launched ballistic missiles, and a few hundred carried by bomber aircraft. These are thermonuclear weapons, considerably more powerful than the fission bombs that devastated Hiroshima and Nagasaki. Looking over these figures, one can hardly help asking, what are all these nuclear weapons for?

---

1. The term "operationally deployed" refers to nuclear warheads that are installed on missiles ready to be fired plus bombs that are ready to be loaded on bombers in service.

There was a rationale for maintaining a very large nuclear arsenal during the cold war: we had to be sure that the Soviets would be deterred from a surprise attack on the United States by their certainty that enough of our arsenal would survive any such attack to allow us to deliver a devastating response. I don't say that U.S. strategic requirements were actually calculated in this way, but the need for such a deterrent at least provided a rational argument for a large arsenal.

This rationale for a large nuclear arsenal is now obsolete. No country in the world could threaten our submarine-based deterrent, and even with an implausibly rapid development of nuclear weapons and missiles, for decades to come no country except Russia will be able to threaten more than a tiny fraction of our land-based deterrent.

Russia maintains a nuclear arsenal of a size similar to ours, though with a different mix of delivery vehicles. On May 24, 2002, Presidents Bush and Putin signed a treaty calling for a reduction in operationally deployed nuclear weapons on both sides to about 3,800 in 2007 and to about 1,700 to 2,200 in 2012. This treaty will almost certainly be ratified by the Senate; Democrats will generally be glad of any reduction in nuclear arms, and few Republicans will want to oppose President Bush on a matter of foreign relations. President Bush has said, "This treaty will liquidate the legacy of the cold war." But any celebration would be premature, for there is far less to this treaty than meets the eye.

For one thing, the rate of reduction is painfully slow. The START III agreement that was announced (though not signed or ratified) by Presidents Clinton and Yeltsin called for a reduction to about 2,000 to 2,500 "strategically deployed" nuclear weapons by 2007, not by the 2012 deadline of the Bush-Putin treaty. (The term "strategically deployed" differs from "operationally deployed" in including all weapons that are associated with delivery systems, whether or not they are actually ready to fire. Thus, for instance, the nuclear warheads of missiles on a submarine in dry-dock would be included on the list of strategically deployed weapons but not of operationally deployed weapons. When this difference is taken into account, the limit of 2,000 to 2,500 strategically deployed weapons in 2007 set by the START III agreement is the same as the limit of 1,700 to 2,200 operationally deployed nuclear missiles in 2012 set by the Bush-Putin treaty.) The treaty is highly reversible; either party can withdraw with forty-five days' notice, and unless renewed the treaty will expire in 2012. Also, unlike former arms control agreements, the Bush-Putin treaty would not call for the destruction of missiles or bombers, only for the removal of their nuclear warheads or bombs.

Most important, the Bush administration has turned back Russian efforts to require in this treaty that the nuclear weapons withdrawn from deployed missiles and bombers should be destroyed. The Defense Department's plans for nuclear weapons have been laid out in a classified *Nuclear Posture Review*,[2] dated January 9, 2002, of which large sections were leaked a few months later to the press.[3] The plans laid out in this review call for the retention of about seven thousand intact warheads that are not operationally deployed, not to mention a large number of plutonium "pits" (the fission bomb that triggers a thermonuclear explosion) and other weapon components. Of course, the treaty does not call for the destruction of demobilized Russian nuclear weapons either, and it does not constrain Russian nuclear tactical weapons, so it actually increases the danger that some Russian weapons could fall into the hands of rogue states or terrorists. This treaty is not designed to "liquidate the legacy of the cold war," as President Bush claims,

---

2. Nuclear Posture Reviews are reports on U.S. nuclear capabilities and plans requested by Congress from the Department of Defense. There have been just two of these reviews; the first was prepared during the Clinton administration.

3. The contents of the *Nuclear Posture Review* were first reported by William Arkin in the *Los Angeles Times* on March 10, 2002. The leaked version is available at www.globalsecurity.org/wmd/library/policy/dod/npr.htm. There is also an unclassified briefing on the *Nuclear Posture Review* by J. D. Crouch, Assistant Secretary of Defense for International Security Policy, available at www.defenselink.mil/news/Jan-2002/t01092002_t0109npr.html.

but to hold on to as much of that legacy as possible. Even taking into account the reductions called for by the Bush-Putin treaty, we are left still wondering what all these nuclear weapons are for.

There is one possible use of a large American nuclear arsenal: to launch a preemptive attack against Russian strategic weapons. I say "Russian," because the large size of our arsenal would be irrelevant for a preemptive attack against any other power. If we ever found that a hostile "rogue" state were about to deploy a few dozen nuclear-armed intercontinental ballistic missiles, and if we could locate them, then they could be destroyed by only a tiny fraction of our nuclear arsenal, and in fact even by conventionally armed cruise missiles. On the other hand, even though we were unable to neutralize the Soviet deterrent during the cold war, now as Russian nuclear forces become increasingly immobile, with their missile-launching submarines tied up at dockside and their land-based mobile intercontinental ballistic missiles kept in fixed garrisons, our large nuclear arsenal may put Russian nuclear forces at risk of being largely destroyed by a preemptive U.S. strike. In the letter of transmittal of the *Nuclear Posture Review* to Congress, Secretary Rumsfeld said that "the US will no longer plan, size, or sustain its forces as though Russia presented merely a smaller version of the threat posed by the former Soviet Union." But that appears to be just what we are doing.

It might seem that the ability to launch a preemptive strike against Russian strategic nuclear forces is a good one to have, but in fact it poses enormous dangers, and to us as well as to Russia. The Russians can count missiles as well as we can, and as "prudent" defense planners, they are likely to rate our chances of a successful preemptive attack more highly than we would. Even though they may understand that the United States now has no plans for such a preemptive attack, they are bound to consider the possibility that this could change if relations between Russia and the United States sour in future. This possibility is likely to seem more probable if the United States proceeds with a national missile defense, which might be perceived to have some effectiveness against a ragged Russian second strike, or if we follow the recommendation of the *Nuclear Posture Review* that the United States should develop real-time intelligence capabilities of a sort that would allow us to target even mobile Russian missiles on the road.

The danger is not that the Russians will get angry with us, or plan to attack us. The danger is that they will quietly adopt a cheap and easy defense against a preemptive American attack, by keeping their forces on a hair-trigger alert. This presents the US with the threat of a large-scale Russian attack by mistake during some future crisis; for instance, the Russians may receive misleading warnings of an imminent American attack and launch their own nuclear weapons before they can be destroyed on the ground. (According to Russian sources, it now takes fifteen seconds for the Russians to target their intercontinental ballistic missiles, and then two to three minutes to carry out the launch.) This danger is exacerbated by the gradual decay of Russia's capabilities for surveillance of possible attacks and control of their own forces, a decay that has already led them on one occasion to mistake a U.S. research rocket launched from a Norwegian island for an offensive missile launched from an American submarine in the Norwegian Sea.[4]

For those who may think that this is a paranoid worry, perhaps left over from cold war movies like *Fail Safe* or *Dr. Strangelove*, it is instructive to look back at mistakes made by American strategic forces during the Cuban missile crisis, the most dangerous crisis of the cold war:[5]

(1) On August 23, 1962, a navigational error led a B-52 bomber on airborne alert—i.e., ready to retaliate if the United States were attacked—that was supposed to be on a nonprovocative course heading over the Arctic Ocean toward Alaska to head instead directly toward the Soviet

---

4. Apparently the Russians were informed in advance of the launch, but somehow this notice did not reach their strategic control center.

5. These examples are taken from Scott Sagan, *The Limits of Safety: Organizations, Accidents, and Nuclear Weapons* (Princeton University Press, 1993).

Union. Its error was noticed when the bomber was only three hundred miles away from Soviet airspace. Despite this incident, and the well-known difficulties of navigation above the Arctic Circle, the routes of U.S. bombers on airborne alert were not changed for months, not until after the October missile crisis. Luckily no similar navigational errors were made by our bombers during the missile crisis.

(2) On October 26, 1962, when United States and Soviet forces were already at a heightened state of alert, an intercontinental ballistic missile was launched from Vandenberg Air Force Base, as part of a test program that no one had thought to cancel. We do not know if the Soviets detected this launch, but they might have.

(3) The Cuban missile crisis happened to come at a time when new Minuteman I missiles were being installed at Malmstrom Air Force Base in Montana. In order to get these missiles ready for possible launch, Air Force and contractor personnel apparently bypassed safeguards that had been designed to prevent a launch by a single officer. Fortunately no officer decided to launch the missiles under his control.

We don't know what mistakes may have been made on the Soviet side. Whatever mistakes were made on either side did not lead to war, but this was evidently not because national leaders are able to completely control their forces under crisis conditions. As President Kennedy said during the Cuban missile crisis, "There is always some son-of-a-bitch who doesn't get the word."

Even though the threat of a large Russian mistaken attack is not acute, it is chronic. It is also the only threat we face that could destroy our country beyond our ability to recover. Compared with this threat, all other concerns about terrorism or rogue countries shrink into insignificance.

This brings me to the one real value of our large nuclear arsenal: we can trade away most of our arsenal for corresponding cuts in Russian forces. I don't mean cuts to about two thousand deployed weapons, but to not more than a few hundred deployed weapons on each side, and with each side having not more than a thousand nuclear weapons of all sorts, including those in various reserves, as called for by a 1997 report of the Committee on International Security and Arms Control of the National Academy of Sciences.[6] In that way, although the danger of a mistaken Russian launch would not be eliminated, the stakes would be millions or tens of millions of casualties, not hundreds of millions.

Such cuts would also reduce the danger that Russian nuclear weapons or weapons material could be diverted to criminals or terrorists. Instead of seeking the maximum future flexibility for both sides in strategic agreements with the Russians, we should be seeking the greatest possible irreversibility on both sides, based on binding ratified treaties. We ought also to be spending more on the program, originally sponsored by former Senator Sam Nunn and Senator Richard Lugar, that assists the Russians in controlling or destroying their excess nuclear materials. At this moment, when the Russians are eager to improve relations with the West, when considerations of economics provide them with a powerful incentive to reduce their nuclear forces, and when for the first time they have a president powerful enough to push such reductions through their military and political establishments, we have an unprecedented opportunity to begin to escape from the risk of nuclear annihilation. It is tragic that we are letting this opportunity slip away from us.

Not only are we not moving fast or far enough in the right direction—in some respects the Bush administration seems to be moving in just the wrong directions. One example is the abrogation of the 1972 treaty limiting antiballistic missile systems.[7] Another example is the revival of the idea of developing nuclear weapons for use, rather than solely for deterrence.

For instance, the *Nuclear Posture Review* calls for the develop-

6. *The Future of US Nuclear Weapons Policy* (National Academy Press, 1997). More recently, a statement by Hans Bethe, Richard Garwin, Marvin Goldberger, Kurt Gottfried, Walter Kohn, and myself has called for an accelerating reduction of our nuclear arsenal and other steps to improve our security; see www.ucsusa.org.

7. See Chapter 7 in this collection.

ment of low-yield, earth-penetrating nuclear weapons for attacks on underground facilities, such as biological warfare laboratories in countries like Iraq. There are great technical difficulties here, which might prevent our using such a weapon even if we had it. When dropped from a bomber, our present earth-penetrating weapon, the B61-11, has penetrated only about ten feet into frozen tundra. The depth of penetration can be increased by accelerating the weapon down to the surface with a rocket; but increasing the velocity of impact beyond a certain point just causes the weapon to crumple, so that instead of the depth of penetration increasing, it decreases. Recent calculations show that an earth-penetrating weapon cannot be driven down to a depth greater than about four times its length in concrete.[8] This sets an upper limit on the depth of penetration of about eighty feet for a weapon that is twice the length of our B61-11. The actual depth that may be reached in practice may be considerably less, because the velocity of impact must be kept low enough to preserve the weapon's electrical circuits.

It is true that an eighty-foot depth is sufficient to put most of the energy of the explosion into a destructive underground blast wave, which can destroy facilities below the actual explosion, but even so, a one-kiloton explosion would only destroy tunnels that are at depths considerably less than three hundred feet, and not much more than that in a horizontal direction; the precise ranges are sensitive to geological details that we are not likely to know. An earth-penetrating nuclear weapon would be effective only against an underground target that is not too deep, and whose location is accurately known. To have confidence that the underground target had been destroyed we would have to have troops on the ground anyway, so that a nuclear attack might not even be necessary.

Even if an earth-penetrating nuclear weapon could destroy its target, we would be unlikely to use it because of radiation effects.

8. Robert W. Nelson, "Low-yield Earth-penetrating Nuclear Weapons," *Science and Global Security* 10, no. 1 (January 2002).

In the 1950s a project known as Plowshare exploded a number of nuclear devices at various depths underground, with the hope of developing peaceful uses for nuclear explosions, like digging canals. Experience in these tests showed that to keep a nuclear explosion from breaking through the surface and spreading radioactive dirt into the atmosphere, a one-kiloton explosion would have to be kept below three hundred feet, with the depth required decreasing only slowly as the yield is decreased. (The penetration of a weapon through the earth would create a shaft to the surface, something that did not exist in the Plowshare tests, so the depth required to avoid fallout is bound to be even larger than indicated by these tests.) To avoid fallout from a nuclear explosion at a depth of only eighty feet it would be necessary to reduce the yield to nineteen tons of TNT, not much more than could be delivered using conventional explosives. I don't believe that there is any way for a nuclear weapon with a yield greater than a few tenths of a kiloton to penetrate to depths sufficient to avoid producing a great deal of radioactive fallout, without someone carrying it down in an elevator.

The fallout produced by a one-kiloton explosion at a depth of eighty feet would kill everyone on the surface within a radius of about half a mile. This estimate is for fallout under conditions of still air; wind could carry the fallout for tens of miles. We could be killing not only the local population, which (as in Afghanistan) we might be trying to enlist on our side, but also whatever forces we or our allies have on the ground.

There was another sign of increased interest in developing nuclear weapons for actual use in a recent statement by William Schneider, the chairman of the Defense Science Board. He announced a renewed study of nuclear-armed interceptor missiles as part of a system of missile defense. Nuclear-armed missile defense interceptors would have technical and political problems of their own, problems that have led to the abandonment of nuclear-armed interceptors as components in missile defense since the administration of Ronald Reagan.

For the dubious advantages of developing new nuclear weapons, we would pay a high price, including pressure for resumed testing of nuclear weapons. As I mentioned, calculations indicate that any nuclear weapon that would be effective against underground targets would release large quantities of radioactivity. Even if the depth of penetration of a nuclear weapon were somehow increased and the yield decreased enough so that no fallout was expected, how, without testing these weapons in action, could anyone ever have confidence that fallout would not escape, especially after the U.S. weapon has created its own shaft to the surface? And how could anyone have confidence in a missile defense system based on nuclear-armed interceptors without tests that involve nuclear explosions in or above the atmosphere? We have not carried out even underground tests since the previous Bush administration. And, as is very much in our interest, neither has Russia or China.

The development of new nuclear weapons for war-fighting would in itself violate our commitment under the 1970 Nuclear Nonproliferation Treaty to deemphasize the role of nuclear weapons and to work toward their total elimination. The resumption of nuclear testing for this purpose would make this violation concrete and dramatic, and would thereby gravely undermine the effectiveness of the Nonproliferation Treaty in discouraging nuclear weapons programs throughout the world.

A special danger of programs to develop nuclear weapons for use is that they may stand in the way of a really large-scale mutual reduction of nuclear arms. I'm not sure whether we are retaining a huge nuclear arsenal in order to facilitate such new weapons programs, or whether the weapons programs are being proposed in order to slow down cuts in our nuclear arsenal. Probably something of both. Back in the days when the first test ban treaty was being debated, one of the arguments against it was that it would stand in the way of Project Plowshare and also Project Orion, the development of a spacecraft propelled by nuclear explosions. (Both programs have long since been abandoned.) The development of nuclear weapons for attacking underground facilities or for missile defense may be today's Orion and Plowshare.

But the current proposals for new nuclear weapons are much more dangerous than the Plowshare or Orion programs. As the world's leader in conventional weaponry, we have a very strong interest in preserving the taboo against the use of nuclear weapons that has survived since 1945. Developing and testing new nuclear weapons for actual use rather than deterrence teaches the world a lesson that nuclear weapons are a good thing to have. This is not entirely a rational matter. I remember that once in the late 1960s I had lunch at MIT with a senior scientific adviser to the government of India. I asked about India's plans for developing and testing nuclear weapons, and he said that it all depended on whether the United States and the Soviet Union could reach an agreement banning all future nuclear testing. I said that that seemed irrational, because it was not the United States or the Soviet Union that presented a military threat to India, and even if such a threat did develop, American and Soviet nuclear forces would in any case be so much greater than India's that it would not matter to India if the United States or the Soviet Union had stopped testing or gone on testing.

The Indian science adviser answered that politics is not always based on rational calculations, that there was great political dissension in Indian governing circles over whether to develop nuclear weapons, and that the spectacle of continued testing of nuclear weapons by the United States or the Soviet Union would strengthen the hands of those in India who favored developing nuclear weapons. Of course, the United States and the Soviet Union did not stop testing at that time; India did develop nuclear weapons; and Pakistan followed suit. Is it likely that resumed U.S. nuclear testing would have no effect on decisions about nuclear weapons in countries like Japan, or Egypt, or Germany? Is it likely that the Nonproliferation Treaty will survive when the United States is developing and testing nuclear weapons for actual use?

After the signing of the Bush-Putin treaty, President Bush was asked why it was necessary for us to keep two thousand nuclear weapons loaded on missiles. He answered that the future was uncertain. The same argument is often made to defend the development of new nuclear weapons. It is true that the future is uncertain, but where is it written that the way to reduce uncertainty is always to maximize our nuclear capabilities? We cannot tell what crisis may occur in U.S.-Russian relations, a crisis that could put the United States at risk from a mistaken launch on their part. We cannot tell what terrorists may take over or steal part of the Russian arsenal. We cannot tell what dangers we may face from a large Chinese arsenal, built to preserve their deterrent from the threat of an American first strike backed up by a missile defense system. We cannot tell what countries may be tipped toward a decision to develop nuclear weapons by new U.S. weapons programs or resumed nuclear testing. There is no certainty whatever we do. We have to decide what are the most important dangers, and these dangers may be increased rather than decreased by other countries' responses to our own weapons programs. The *Nuclear Posture Review* strikingly fails to consider what other countries might do in response to our plans for nuclear weapons.

At the beginning of the twentieth century, Britain was overwhelmingly the world's greatest naval power, much as the United States is today the world's leader in conventional arms. Then in 1905 Admiral Sir John Fisher, the First Sea Lord, pushed forward the construction of a new design for a fast battleship with a main armament consisting solely of twelve-inch guns, the biggest guns then available. The prototype was named the *Dreadnought,* a name previously used in the Royal Navy for seven other warships, going back to the sixteenth century. Dreadnoughts really were superior to all previous battleships, and suddenly what counted was not the size of a country's fleet, in which Britain was supreme, but the number of its dreadnoughts. Other countries could now compete with Britain by building dreadnoughts, and a naval arms race began between Britain and Germany, in which Britain would stay ahead only with great expense and difficulty. Admiral of the Fleet Sir

Frederick Richards complained in Parliament that "the whole British fleet was morally scrapped and labeled obsolete at the moment when it was at the zenith of its efficiency and equal not to two but practically to all the other navies of the world combined."[9] Like dreadnoughts, nuclear weapons can act as an equalizer between strong nations like the United States, with great economic and conventional military power, and weaker countries or even terrorist organizations. It should be clear by now that national security is not always best served by building the best weapons.

As a scientist, I can recognize a kind of technological restlessness at work, from the building of the *Dreadnought* to this year's *Nuclear Posture Review.* Years before he pioneered the *Dreadnought,* as a newly appointed captain in charge of the Royal Navy's torpedo school, Fisher explained that "if you are a gunnery man, you must believe and teach that the world is saved by gunnery, and will only be saved by gunnery. If you are a torpedo man, you must lecture and teach the same thing about torpedoes." There is nothing corrupt or unpatriotic about such attitudes, but their consequences could be catastrophic.

———

Since this article was published, plans for earth-penetrating nuclear weapons for attacks on underground facilities seem to have been shelved. In October 2005 the Bush administration dropped its request for funding of this weapon.

It is difficult to find much else in the way of encouragement. Relations between Russia and the United States have become more hostile, in part because of the expansion of NATO and U.S. plans for missile defense deployments in states of the former Warsaw pact. Meanwhile, Russia is playing a more aggressive role in the Caucasus and elsewhere. Above all, both Russia and the United States remain threatened by the enormous nuclear arsenals of both

9. Quotes of Richards and Fisher are taken from Robert K. Massie, *Dreadnought: Britain, Germany, and the Coming of the Great War* (Ballantine, 1991).

countries. In an article published by the American Philosophical Society in March 2008, Richard Garwin remarked: "I judge that luck played a very major role in the avoidance of nuclear war in the 1960s and 1970s and that an all-out nuclear war could still take place by accident."

9

Is the Universe a Computer?

I was not enthusiastic when I heard of a new book called *A New Kind of Science* (Wolfram Media, 2002), by a computer whiz, Stephen Wolfram, who had left particle physics. As I gathered from advertisements, Wolfram was proposing to abandon the search for quantum field theories or string theories that would tie up all the loose ends in fundamental physics—the sort of physics I do—in favor of a vision of the universe as something like a gigantic computer. Most of my writing for the *New York Review of Books* has not been reviews of books, but when Robert Silvers asked me to review Wolfram's book, I felt that I could not pass up the opportunity to defend the old kind of science.

As it happened, I found a good deal that was interesting in Wolfram's book. Its claims are overblown, but it suggests intriguing directions for future research in computer science. My review was published by the *New York Review of Books* in October 2002.

Everyone knows that electronic computers have enormously helped the work of science. Some scientists have had a grander vision of the importance of the computer. They expect that it will change our view of science itself, of what it is that scientific theories are supposed to accomplish, and of the kinds of theories that might achieve these goals.

I have never shared this vision. For me, the modern computer is only a faster, cheaper, and more reliable version of the teams of clerical workers (then called "computers") that were programmed at Los Alamos during World War II to do numerical calculations. But neither I nor most of the other theoretical physicists of my generation learned as students to use electronic computers. That skill was mostly needed for number crunching by experimentalists who had to process huge quantities of numerical data, and by theorists who worked on problems like stellar structure or bomb design. Computers generally weren't needed by theorists like me, whose work aimed at inventing theories and calculating what these theories predict only in the easiest cases.

Still, from time to time I have needed to find the numerical solution of a differential equation,[1] and with some embarrassment I would have to recruit a colleague or a graduate student to do the job for me. It was therefore a happy day for me when I learned to use a program called Mathematica, written for personal computers under the direction of Stephen Wolfram. All one had to do was to type out the equations to be solved in the prescribed code, press shift-enter, and, presto, the answer would pop up on the monitor screen. The Mathematica user's manual now sits on my desk, so fat and heavy that it does double duty as a bookend for the other books I keep close at hand.

Now Wolfram has written another book that is almost as heavy as the Mathematica user's manual, and that has attracted much attention in the press. *A New Kind of Science* describes a radical vision of the future of science, based on Wolfram's long love affair with computers. The book's publisher, Wolfram Media, announces "a whole new way of looking at the operation of our universe" and "a series of dramatic discoveries never before made public." Wolfram claims to offer a revolution in the nature of science, again

---

1. A differential equation gives a relation between the value of some varying quantity and the rate at which that quantity is changing, and perhaps the rate at which that rate is changing, and so on. The numerical solution of a differential equation is a table of values of the varying quantity, that to a good approximation satisfy both the differential equation and some given conditions on the initial values of this quantity and of its rates of change.

and again distancing his work from what he calls traditional science, with remarks like "If traditional science was our only guide, then at this point we would probably be quite stuck." He stakes his claim in the first few lines of the book: "Three centuries ago science was transformed by the dramatic new idea that rules based on mathematical equations could be used to describe the natural world. My purpose in this book is to initiate another such transformation."

Usually I put books that make claims like these on the crackpot shelf of my office bookcase. In the case of Wolfram's book, that would be a mistake. Wolfram is smart, winner of a MacArthur Fellowship at age twenty-two, and the progenitor of the invaluable Mathematica, and he has lots of stimulating things to say about computers and science. I don't think that his book comes close to meeting his goals or justifying his claims, but if it is a failure it is an interesting one.

The central theme of the book is easily stated. It is that many simple rules can lead to complex behavior. The example that is used repeatedly to illustrate this theme is a favorite toy of complexity theorists known as the cellular automaton, so I will have to say a bit about what cellular automata are.

Take a piece of white graph paper that has been crosshatched into little squares. These are the "cells." Blacken one or more of the cells in the top row, chosen any way you like, leaving all the others white. This is your input. Now blacken some cells in the second row, according to some fixed rule that tells you to make any cell black or leave it white depending on the colors of its three neighboring cells in the first row (that is, the cells in the first row that are either immediately above the cell in the second row or one cell over to the right or left.) Then use the same rule, whatever it is, automatically to color each cell in the third row according to the colors of its three neighboring cells in the second row, and keep going automatically in the same way to the rows below. The coloring rule used in this way is an elementary cellular automaton.

This may seem like a solitaire variation on tic-tac-toe, only not as exciting. Indeed, most of the 256 possible[2] elementary cellular automata of this sort are pretty boring. For instance, consider rule 254, which dictates that a cell is made black if the cell immediately above it, or above it and one space over to the left or right, is black, and otherwise it is left white. Whatever the input pattern of black cells in the top row, the black cells will spread in the rows below, eventually filling out an expanding black triangle, so that the cells in any given column will all be black once you get to a low-enough row.

But wait. Wolfram's prize automaton is number 110 in his list of 256. Rule 110 dictates that a cell in one row is left white if the three neighboring cells in the row above are all black or all white or black-white-white, and otherwise it is made black. After applying this rule twenty or thirty times with a very simple input, in which just one cell is made black in the top row, one sees nothing interesting. Wolfram programmed a computer to run this automaton, and he ran it for millions of steps. After a few hundred steps something surprising happened: the rule began to produce a remarkably rich structure, neither regular nor completely random. A pattern of black cells spreads to the left, with a foamy strip furthest to the left, then a periodic alternation of regions of greater and lesser density of black cells which moves to the right, followed by a jumble of black and white cells. It is a dramatic demonstration of Wolfram's conclusion, that even quite simple rules and inputs can produce complex behavior.

Wolfram is not the first to have worked with cellular automata. They had been studied for decades by a group headed by Edward

2. The automaton must tell you the color of a cell in one row for each of the $2 \times 2 \times 2 = 8$ possible color patterns of the three neighboring cells in the row above, and the number of ways of making these eight independent decisions between two colors is $2^8 = 256$. In the same way, if there were 3 possible colors, then the number of coloring decisions that would have to be specified for each cell by an elementary cellular automaton would be $3 \times 3 \times 3 = 27$, and the number of automata (calculated using Mathematica) would therefore be $3^{27} = 7625597484987$.

Fredkin at MIT, following the ground-breaking work of John Von Neumann and Stanislas Ulam in the 1950s. Wolfram is also not the first to have seen complexity coming out of simple rules in automata or elsewhere. Around 1970 the Princeton mathematician John Horton Conway invented "The Game of Life," a two-dimensional cellular automaton in which cells are blackened according to a rule depending on the colors of all the surrounding cells, not just the cells in the row above. Running the game produces a variety of proliferating structures reminiscent of micro-organisms seen under a microscope. For awhile the Game of Life was dangerously addictive for graduate students in physics. A decade later another mathematician, Benoit Mandelbrot, the inventor of fractals, gave a simple algebraic prescription for constructing the famous Mandelbrot set, a connected two-dimensional figure that shows an unbelievable richness of complex detail when examined at smaller and smaller scales.

There are also well-known examples of complexity emerging from simple rules in the real world. Suppose that a uniform stream of air is flowing in a wind tunnel past some simply shaped obstacle, like a smooth solid ball. If the air speed is sufficiently low, then the air flows in a simple smooth pattern over the surface of the ball. Aerodynamicists call this laminar flow. If the air speed is increased beyond a certain point, vortices of air appear behind the ball, eventually forming a regular trail of vortices called a "Von Karman street." Then as the air speed is increased further, the regularity of the pattern of vortices is lost, and the flow begins to be turbulent. The air flow is then truly complex, yet it emerges from the simple differential equations of aerodynamics and the simple setup of wind flowing past a ball.

What Wolfram has done that seems to be new is to study a huge number of simple automata of all types, looking specifically for those that produce complex structures. There are cellular automata with more than two colors, or with coloring rules like the Game of Life that change the colors of cells in more than one row at a time, or with cells in more than two dimensions. Beyond cellular automata, there are also automata with extra features like memory, including the Turing machine, about which more later. From his explorations of these various automata, Wolfram has found that the patterns they produce fall into four classes. Some are very simple, like the spreading black triangle in the rule 254 elementary cellular automaton that I mentioned first. Other patterns are repetitive, such as nested patterns that repeat themselves endlessly at larger and larger scales. Still others seem entirely random. Most interesting are automata of the fourth class, of which rule 110 is a paradigm. These automata produce truly complex patterns, neither repetitive nor fully random, with complicated structures appearing here and there in an unpredictable way.

So what does this do for science? The answer depends on why one is interested in complexity, and that depends in turn on why one is interested in science.

Some complex phenomena are studied by scientists because the phenomena themselves are interesting. They may be important to technology, like the turbulent flow of air past an airplane, or directly relevant to our own lives, like memory, or just so lovely or strange that we can't help being interested in them, like snowflakes. Unfortunately, as far as I can tell, there is not one real-world complex phenomenon that has been convincingly explained by Wolfram's computer experiments.

Take snowflakes. Wolfram has found cellular automata in which each step corresponds to the gain or loss of water molecules on the circumference of a growing snowflake. After adding a few hundred molecules some of these automata produce patterns that do look like real snowflakes. The trouble is that real snowflakes don't contain a few hundred water molecules, but more than a thousand billion billion molecules. If Wolfram knows what pattern his cellular automaton would produce if it ran long enough to add that many water molecules, he does not say so.

Or take complex systems in biology, like the human nervous or immune systems. Wolfram proposes that the complexity of such systems is not built up gradually in a complicated evolutionary history, but is rather a consequence of some unknown simple rules, more or less in the way that the complex behavior of the pattern produced by cellular automaton 110 is a consequence of its simple rules. Maybe so, but there is no evidence for this. In any case, even if Wolfram's speculation were correct it would not mean that the complexity of biological systems has little to do with Darwinian evolution, as Wolfram contends. We would still have to ask why organisms obey some simple rules and not other rules, and the only possible answer would be that natural selection favors those rules that generate the kind of complex systems that improve reproductive fitness.

Wolfram even tackles the old conflict between belief in a deterministic view of nature and in the existence of free will. He suggests that free will is an illusion that arises from the apparent unpredictability of the complex behavior produced by those simple rules of biology that he imagines to govern the human organism. This is odd, because we certainly don't attribute free will to other unpredictable complex phenomena like earthquakes or thunderstorms. Surely the impression of free will arises instead from our personal experience of actually deciding what to do, an experience that we are unwilling to suppose is enjoyed by earthquakes or thunderstorms. This is not to say that I have any enlightenment to offer about free will either, because I have never been able to understand the inconsistency that other people find between free will and a completely deterministic view of nature. Free will to me means only that we sometimes decide what we do, and we know that this is true by the same sort of mental experience that convinced Descartes that he existed, but we have no mental experience that tells us that what we want to do is not an inevitable consequence of past conditions and the laws of nature.

Other scientists like myself study phenomena that may not be intrinsically very interesting, because they think that studying these phenomena will help them to understand the laws of nature, the rules that govern all phenomena. In such work, we tend to study the simplest possible phenomena, because it is in these cases that we can most easily calculate what our theories predict, and compare the results with experimental data to decide whether our theories are right or wrong. Wolfram makes it seem that physicists choose simple rather than complex phenomena to study because of long habit or mathematical flabbiness, but in seeking the laws of nature it is the essence of the art of science to avoid complexity.

My own work has been mostly on the theory of elementary particles, but I have never found these particles very interesting in themselves. An electron is featureless, and much like every other electron. Most of those of us who study elementary particles do so because we think that at this moment in the history of science it is the best way to discover the laws that govern all natural phenomena. Because of this special motivation, we don't generally care whether we can calculate everything that happens to elementary particles in complicated situations, only whether we can calculate enough to check the validity of our theories. In collisions of elementary particles at moderate energy the energy of the collision goes into complex showers of particles. No one can predict the details of these showers, even where the underlying theory is known, and hardly anyone cares. At higher energies things are simpler: the energy goes into well-defined jets, each containing particles that travel in the same direction, in a way that can be calculated theoretically and compared with experiment to test our theories of elementary particles. It is the laws, not the phenomena, that interest us.

Unlike elementary particles, planets have historically seemed interesting for religious and astrological reasons. But it was the *simplicity* of planetary motions that allowed Newton to discover the laws of motion and gravitation. The planets move in empty space and to a good approximation under the influence of a single motionless body, the Sun. Newton would never have discovered his laws by studying turbulence or snowflakes.

Wolfram himself is a lapsed elementary particle physicist, and I suppose he can't resist trying to apply his experience with digital computer programs to the laws of nature. This has led him to the view (also considered in a 1981 article by Richard Feynman) that nature is discrete rather than continuous. He suggests that space consists of a network of isolated points, like cells in a cellular automaton, and that even time flows in discrete steps. Following an idea of Edward Fredkin, he concludes that the universe itself would then be an automaton, like a giant computer. It's possible, but I can't see any motivation for these speculations, except that this is the sort of system that Wolfram and others have become used to in their work on computers. So might a carpenter, looking at the moon, suppose that it is made of wood.

There is another reason for studying complex phenomena—not because the phenomena are interesting, which they sometimes are, or because studying complex phenomena is a good way to learn the laws of nature, which it isn't, but because complexity itself is interesting. Maybe there is a theory of complexity waiting to be discovered, that says simple things about complex behavior in general, not just about the rule 110 cellular automaton or turbulent air flow or the human nervous system.

There are other examples of what I like to call free-floating theories, theories that are applicable in a wide (though not unlimited) variety of very different contexts. The theory of chaos, which has captured the public imagination, deals with systems, from the weather to the pebbles in the rings of Saturn, whose behavior exhibits an exquisite sensitivity to initial conditions. Thermodynamics, the science of heat, is a less trendy example. Concepts of thermodynamics like temperature and entropy are applicable to black holes as well as to steam boilers. A less familiar example is the theory of broken symmetry. Many very different substances, including superconductors, magnetized iron, and liquid helium, are governed by equations that have some symmetry, in the sense that the equations look the same from certain different points of view, and yet the substances exhibit phenomena that do not respect this symmetry.

There is a low-intensity culture war going on between scientists who specialize in free-floating theories of this sort and those (mostly particle physicists) who pursue the old reductionist dream of finding laws of nature that are not explained by anything else, but that lie at the roots of all chains of explanation. The conflict usually comes to public attention only when particle physicists are trying to get funding for a large new accelerator. Their opponents are exasperated when they hear talk about particle physicists searching for the fundamental laws of nature. They argue that the theories of heat or chaos or complexity or broken symmetry are equally fundamental, because the general principles of these theories do not depend on what kind of particles make up the systems to which they are applied. In return, particle physicists like me point out that, although these free-floating theories are interesting and important, they are not truly fundamental, because they may or may not apply to a given system; to justify applying one of these theories in a given context you have to be able to deduce the axioms of the theory in that context from the really fundamental laws of nature.

This debate is unfortunate, for both kinds of science are valuable, and they often have much to teach each other. My own work in elementary particle physics has benefited tremendously from the idea of broken symmetry, which originated in the study of the solid state but turned out to be the key both to understanding reactions involving particles called pi mesons at low energy and to the unification of some of the forces acting on elementary particles. The theory of complexity might also have lessons for elementary particle theory (or vice versa) but it is not likely to be fundamental in the same sense as elementary particle physics.

Lately particle physicists have been having trouble holding up their end of this debate. Progress toward a fundamental theory has been painfully slow for decades, largely because the great success of the "Standard Model" developed in the 1960s and 1970s has left us with fewer puzzles that could point to our next step. Scientists studying chaos and complexity also like to emphasize that their work is applicable to the rich variety of everyday life, where elementary particle physics has no direct relevance.

Scientists studying complexity are particularly exuberant these days. Some of them discover surprising similarities in the properties of very different complex phenomena, including stock market-fluctuations, collapsing sand piles, and earthquakes. Often these phenomena are studied by simulating them with cellular automata, such as Conway's Game of Life. This work is typically done in university physics departments and in the interdisciplinary Santa Fe Institute. Other scientists who call themselves complexity theorists work in university departments of computer science and mathematics and study the way that the number of steps in a computer calculation of the behavior of various systems increases with the size of the systems, often using automata like the Turing machine as specific examples of computers. Some of the systems they study, such as the World Wide Web, are quite complex. But all this work has not come together in a general theory of complexity. No one knows how to judge which complex systems share the properties of other systems, or how in general to characterize what kinds of complexity make it extremely difficult to calculate the behavior of some large systems and not others. The scientists who work on these two different types of problem don't even seem to communicate very well with each other. Particle physicists like to say that the theory of complexity is the most exciting new thing in science in a generation, except that it has the one disadvantage of not existing.

It is here I think that Wolfram's book may make a useful contribution. Wolfram and his coworkers have been able to show that numerous simple "class four" automata that produce complex behavior, like the rule 110 cellular automaton, are able to emulate each other. That is, by setting up a suitable input pattern of black and white cells in the rule 110 cellular automaton, one can produce the same complex pattern that would be produced by other class four automata, and vice versa. (In this emulation, blocks of cells in one automaton represent a single cell in the automaton being emulated.) What makes this particularly interesting is that one of the automata that can be emulated in this way is the universal Turing machine.[3]

The Turing machine is the most important automaton in the history of computer science, and the forerunner of today's digital computers. It was invented in 1936 by Alan Turing, who in World War II became one of Britain's ace cipher-breakers and was later the hero of Hugh Whitemore's play *Breaking the Code*. Turing's purpose was to answer a classic question of mathematical logic known as the Decision Problem: given some deductive mathematical system like arithmetic or Euclidean geometry or symbolic logic, is there any logical method that, when applied mechanically to any statement of that system, is guaranteed to decide whether that statement can be proved by following the rules of that system?[4]

To answer this question, the Turing machine was designed to capture the essence of mechanical logical methods. Just as a person going through a mathematical proof works with a string of symbols, focusing on just one at a time, the Turing machine works on a one-dimensional sequence of cells, each containing a symbol taken from some finite list, with only one "active" cell that can be read and possibly changed at each step. Also, to correspond to the fact that a person working out a proof would keep some memory

3. Wolfram says that the main elements of the proof were found in 1994 by one of his assistants, Matthew Cook, and he gives an unreadable updated version in this book, along with an admission that a few errors may still remain. I gather that the proof has not been published in a refereed journal. An article in *Nature* by Jim Giles titled "What Kind of Science Is This?" (May 16, 2002) reports that when Cook left his job with Wolfram in 1998, he gave a talk on his work at the Santa Fe Institute, but the talk did not appear in the conference proceedings; Wolfram took legal action against Cook, arguing that Cook was in breach of agreements that prevented him from publishing until after the publication of Wolfram's book.

4. Turing took pains to point out that this issue had not been settled by the famous 1931 theorem of Kurt Gödel, which states that there are statements in the general system of mathematics presented in the *Principia Mathematica* of Bertrand Russell and Alfred North Whitehead that can be neither proved nor disproved by following the rules of that system.

of previous steps, Turing gave his machine a memory register, which can be in any one of a finite number of "conditions."

Each type of Turing machine obeys a fixed rule that tells it at each step how to change the symbol in the active cell, how to change the condition of the memory register, and whether to move the active cell one step to the left or right, according to the symbol in the active cell and the condition of the memory register. It makes no difference what symbols we choose to use or what conditions are possible for the memory register; their significance arises only from the rules of the machine. The problem to be solved and the data to be used are fed into the machine as an initial string of symbols, and the answer appears as the string of symbols found when the memory register reaches a condition that tells the machine to stop.

Turing never actually built such a machine (though he did go on to build some special-purpose computers), but if you like you can think of the cells in a Turing machine as forming a paper tape, with the symbols just a sequence of colored dots on the tape, read and written by a scanning device that moves up or down the tape from one active cell to another. In this example the memory register is a simple mechanical pointer which can take any of two or more positions. The decision how to change the color of the active cell and the position of the memory register and how to move the tape is made according to the color of the cell being read and the position of the pointer, following rules that are hard-wired into the machine. A specific Turing machine is entirely characterized by the number of possible colors on the cells of the tape, the number of possible positions of the pointer, and by the rules wired into the machine.

The important thing about Turing machines is that some of them are *universal*. Turing was able to prove that any of these universal Turing machines could calculate or prove anything that could be calculated or proved by any other Turing machine. For instance, at least one of the huge number (96 to the power 48) of the possible Turing machines that use two colors and have a memory register with twenty-four positions is universal. Further, from the way that Turing machines were designed to imitate the way humans mechanically do calculations, Turing argued that universal Turing machines could calculate or prove anything that could be calculated or proved by any purely mechanical procedure. This is often called the Church-Turing thesis, because at about the same time the Princeton mathematician Alonzo Church reached similar conclusions about a more abstract but equivalent mathematical method of his own. Incidentally, Turing and Church found that the answer to the Decision Problem in most mathematical or logical systems is "no": there is no mechanical procedure that is guaranteed to decide whether any given statement can be proved by following the rules of that system.

Since universal Turing machines can be emulated by the rule 110 cellular automaton, it follows that this cellular automaton, and all the other automata that can emulate it, are also universal— they can do any computation that can be done by any computer. The program for the calculation and the data to be used would be fed into a rule 110 cellular automaton as a pattern of black cells in the top row, and the answer would appear as a pattern on a lower row. Wolfram says that all these automata are computationally equivalent, but that is only true in a limited sense. The simpler the design of a universal computer, the more steps it takes to emulate each single step of a practical computer. This is why Dell and Compaq do not sell Turing machines or rule 110 cellular automata.

Wolfram goes on to make a far-reaching conjecture, that almost all automata of any sort that produce complex structures can be emulated by any one of them, so they are all equivalent in Wolfram's sense, and they are all universal. This doesn't mean that these automata are computationally equivalent (even in Wolfram's sense) to systems involving quantities that vary continuously. Only if Wolfram were right that neither space nor time nor anything else is truly continuous (which is a separate issue) would the Turing machine or the rule 110 cellular automaton be computationally equivalent to an analog computer or a quantum computer or a brain or the universe. But even without this far-reaching (and far-out) assumption, Wolfram's conjecture about the computational equivalence of automata would at least provide a starting point for a theory of any sort of complexity that can be produced by any kind of automaton.

The trouble with Wolfram's conjecture is not only that it has not been proved—a deeper trouble is that it has not even been stated in a form that *could* be proved. What does Wolfram mean by complex? If his conjecture is not to be a tautology, then we must have some definition of complex behavior independent of the notion of universality. The pattern produced by the rule 110 cellular automaton certainly looks complex, but what criterion for complexity can we use that would tell us that it is complex enough for Wolfram's conjecture to apply?

There is a well-known parallel problem in defining randomness. The most common precise definition of the randomness of a string of digits or of a sequence of black and white cells on a tape is that it is random if there is no way of describing it with a string of shorter length. The trouble is that according to this definition the string of digits in a number like the square root of two would not qualify as random, because it can be described very simply—it is the square root of two—even though it surely looks random. (Mathematica gives the first thirty digits as 1.41421356237309504880168872420.) In the same way, it would not do to define the output of a cellular automaton like rule 110 with a single black cell in the top row as complex only if it can't be described in simple terms, because it can be described in simple terms—it is the output of rule 110 with a single black cell in the top row. There are other definitions of randomness, such as the absence of correlations: the digits in the square root of two can be said to be random because, as far as is known, being given one digit at an unidentified decimal place tells you nothing at all about what the next digit is likely to be. Wolfram has not even begun to formulate a similar definition of complexity.

In fact, as he admits, for Wolfram the real test of the complexity of a pattern is that it should look complex. Much of his discussion of complexity is anecdotal, relying on pictures of the patterns produced by specific automata that he has known. In this, Wolfram is allying himself with one side in the ancient struggle between what (with much oversimplification) one might call cultures of the image and cultures of the word. In our own time it has surfaced in the competition between television and newspapers and between graphical user interfaces and command line interfaces in computer operating systems.

The culture of images has had the better of it lately. For a while the culture of the word had seemed to have scored a victory with the introduction of sound into motion pictures. In *Sunset Boulevard,* Norma Desmond recalls that in silent films, "We didn't need dialogue. We had faces." But now movies can go on for long stretches with no words, only the thunk of cars running into each other and the sizzle of light sabers. The ascendancy of the culture of the image has been abetted by computers and the study of complexity, which have made possible the simulation of complex visual images.

I am an unreconstructed believer in the importance of the word, or its mathematical analogue, the equation. After looking at hundreds of Wolfram's pictures, I felt like the coal miner in one of the comic sketches in *Beyond the Fringe,* who finds the conversation down in the mines unsatisfying: "It's always just 'Hallo, 'ere's a lump of coal.'"

Wolfram's classification of the patterns produced by cellular automata dates from the early 1980s, and the discovery that the rule 110 elementary cellular automaton is a universal computer was made in the early 1990s. Since then, none of this work has had much of an impact on the research of other scientists, aside from Wolfram's employees. The strongest reaction I have seen by scientists to this new book has been outrage at Wolfram's exaggeration of the importance of his own contributions to the study of complexity. Wolfram's survey of the complex patterns produced by automata may yet attract the attention of other scientists if it leads to some clear and simple mathematical statement about complexity. I doubt if even Wolfram cares what picture is produced by the rule 110 cellular automaton after a billion steps. But if Wolfram can give a precise statement of his conjecture about the computational equivalence of almost all automata that produce complex patterns and prove that it is true, then he will have found a simple common feature of complexity, which would be of real interest. In the study of anything outside human affairs, including the study of complexity, it is only simplicity that can be interesting.

10

Foreword to *A Century of Nature*

Nature is a distinguished British scientific periodical, which publishes research articles in many branches of science along with news of the world of science. Though *Nature* was founded in 1869, it was in the twentieth century that it began to publish research articles of the first importance. In a book published by the University of Chicago Press in 2003, Laura Garwin, formerly the Physical Sciences editor and North American editor of *Nature*, and Tim Lincoln, editor of *Nature*'s News and Views section, brought together reprints of articles reporting twenty-one of the most important discoveries reported in *Nature* in the twentieth century, together with commentaries by some of today's leading scientists. For example, their collection includes Chadwick's report of the 1932 discovery of the neutron, with a commentary by Maurice Goldhaber; Watson and Crick's account of the 1953 discovery of the double helix structure of DNA, with a commentary by Sydney Brenner; and Hewish, Bell, et al.'s report of the 1967 discovery of the pulsar in the Crab Nebula, with commentary by Joseph Taylor. The following is the brief foreword I contributed to Garwin and Lincoln's impressive collection.

Few scientists and even fewer nonscientists will ever have read the original reports of classic scientific discoveries. For instance, in my life I have met only two physicists who have read through Newton's *Principia* (and I am not one of them). And why should they? Science, unlike theology or the arts, is a cumulative enterprise. We now understand things much better than those who preceded us.

Any competent graduate student in theoretical physics today understands Maxwell's theory of electrodynamics better than Maxwell did, so why should he or she read Maxwell's 1873 treatise *Electricity and Magnetism?*

Still, it would be a pity to cut ourselves off completely from our past. The history of science is as interesting as any other branch of history, and for scientists one of the gratifications of our work is the sense of continuing a great historical tradition, of carrying forward the work of our predecessors and preparing the ground for our successors. Indeed, in learning science, we all absorb a certain amount of scientific history: Darwin and Wallace did this, and then Mendel did that, and so on. The trouble with this potted history is that it never captures the extreme difficulty of taking a new step, and it is often just plain wrong. For an understanding of the history of science, there is no substitute for actually reading some of the great works of past scientists. It is therefore cause for celebration that Laura Garwin and Tim Lincoln have assembled this collection of classic scientific papers from *Nature,* with explanatory essays to make the papers accessible to a general audience.

The collection is remarkable, not only for the importance of the individual articles, but also for the fact that they all appeared in the same journal. It used to be assumed that natural philosophers (only later called "scientists") would be able to read and understand reports of new discoveries in any area of science, and could often make contributions to many of them. Think of William Hyde Wollaston, a medical doctor of the early nineteenth century, who discovered two new elements, made the first observations of dark lines in the spectrum of the sun, invented the camera lucida, and gave his name to the premier honor in geology, the Wollaston Medal. Now we are terribly specialized, and can read papers only in our own sub-sub-sub-specialties. Aside from the proceedings of learned academies, there are only two English-language scientific journals that keep up the old tradition of encompassing all of science: *Science* in the United States, and *Nature* in Britain.

In this collection of articles from *Nature* one can read seminal articles in every field of twentieth-century science. Every field, that is, except my own: elementary particle theory. For some reason, particle theorists rarely submit their work to *Science* or *Nature*. (But *Nature* had its chance: Enrico Fermi submitted to *Nature* his great 1932 paper on the theory of beta decay, which founded the modem theory of weak interactions, but it was rejected.) Looking over this collection, I begin to feel that we should think of contributing to *Nature*. We would be in such good company.

11

Ambling toward Apocalypse

The American Academy of Arts and Sciences was founded in Boston in 1780, and since then has contributed greatly to the intellectual life of the Boston area, and, increasingly, of the nation. When my wife and I lived in Cambridge, we often enjoyed the monthly "stated meetings" of the Academy, then at beautiful Brandegee House, at which a talk was followed by dinner and conversation with friends. Since moving to Texas I have had less to do with the Academy, but I've always been glad to have a chance to return there. I was especially glad to be invited to give a talk in honor of two friends, now alas gone from us, who had been presidents of the Academy: Herman Feshbach and Victor Weisskopf. The following is based on the talk I gave on March 12, 2003 at the 1,868th stated meeting of the Academy, by then removed to a stately house of its own. It was published in the Bulletin of the Academy and in the Public Interest Report of the Federation of American Scientists. In this talk I express in another way my concern about the dangers of nuclear terrorism and accidental nuclear war. I have deleted some remarks on missile defense and nuclear weapons development that overlap material in essays 7 and 12 in this collection.

It is always an honor and a pleasure to speak to this Academy, but it is a special honor for me to give a talk dedicated to two great men: Herman Feshbach and Victor Weisskopf. I knew them as senior figures at the Massachusetts Institute of Technology; Viki recruited me to the Physics Department, which he chaired, and

Herman was director of the Center for Theoretical Physics, where I worked. Of course, long before I knew them, they had made their reputations as theoretical physicists. Among other things, both had made major contributions to nuclear physics, which in the 1940s became an important factor in world history. Herman's PhD thesis was on tritium, an isotope that later became an essential ingredient in hydrogen bombs. Viki was one of those at Los Alamos who designed the first atomic bombs, and he felt the heat of the first nuclear explosion at Alamogordo from ten miles away.

The experience of participating in the development of nuclear weapons gave a generation of physicists a sense not of guilt but of responsibility—of what Viki called "an obligation to inform the public about the awesome consequences of a nuclear war . . . our nightmarish vision of an actual nuclear conflict, based on our particular understanding of the power of the weapon we had made." To carry out this aim, Viki and others created the Federation of Atomic Scientists. Later, in 1969, Viki and Herman and I joined thirty-eight other faculty at MIT in forming the Union of Concerned Scientists, of which Herman was the first chairman. In the 1970s Viki worked, through the Academy, at organizing conferences on arms control. During Herman's first term as president of the Academy, the Committee on International Security Studies was established. These organizations have played an essential role in providing the public with independent scientific judgments about national nuclear policy and other matters.

I wish that I could say that with the end of the cold war, these efforts are no longer needed. Unfortunately, the reverse is true. Since September 11, 2001, we have been painfully aware that there are people in the world who hate America so much that they will give their lives to hurt us. If terrorists succeeded in exploding a nuclear weapon in one of our cities, it would kill so many people and do so much damage that it would make September 11 look like an ordinary working day. Given a hundred pounds or so of highly enriched uranium, it would not be difficult to make a nuclear weapon and put it in an American city, on a truck or plane, or in one of the seventeen million containers that freighters bring into North American harbors every year.

Last fall I participated in the Hart-Rudman Independent Task Force on Homeland Security Imperatives, convened by the Council on Foreign Relations. Our task force concluded that "a year after September 11, 2001, America remains dangerously unprepared to prevent and respond to a catastrophic attack on U.S. soil." For instance, as we noted, the American Association of Port Authorities estimates that the cost of adequate physical security at our commercial seaports is about $2 billion, yet only $92.3 million in federal grants had been authorized and approved.

Whatever we do to guard our cities, some vulnerabilities will always remain. We also have to guard against nuclear terrorism by working with other countries to control fissionable materials. Russia now holds about 150 tons of plutonium and 850 tons of highly enriched uranium. Since 1991 the United States has been committed to the Nunn-Lugar Cooperative Threat Reduction Program, which among other things aims to improve Russian control over these materials to keep them out of the hands of terrorists and other states, and to make them unusable for weapons. Our rate of spending on this program, however, is only about a third of what it should be. The planned upgrade of security has been completed for only about 40 percent of Russian nuclear storage sites, and less than a seventh of Russia's stockpile of highly enriched uranium has been made unusable for weapons. Last year President Bush proposed to cut spending on this program by 5 percent; this year he has asked for only about 10 percent in additional funds.

We are not even adequately protecting our own nuclear weapons facilities. Energy Secretary Spencer Abraham has said that his department "is unable to meet the next round of security mission requirements" and has asked for $379.7 million to rectify that situation, but the White House has approved just $26.4 million. There are no technical obstacles here—only a shortage of funds.

One program did receive a flood of new funding after September 11: ballistic missile defense. This past December President

Bush announced the decision to deploy a limited missile defense by October 2004. There is $9 billion in the 2004 budget for missile defense—a figure that will surely increase as the program moves from testing and development to deployment. I have heard estimates that the total cost of the missile defense program through 2014 will reach a trillion dollars.

The irony in the contrast between support for missile defense and for other programs is painful, because attack by ballistic missiles is not only just one of many ways that terrorists could use nuclear weapons against us; *it is the least likely way.* Terrorists may be willing to commit suicide, but the leaders of the states that harbor them never are. Why should anyone attack us with ballistic missiles, which always reveal their source, rather than in any of the many ways that do not?

The real danger is not that a rogue state will launch nuclear-armed intercontinental ballistic missiles at us, but that it will use nuclear weapons in local conflicts or hand them over to terrorists. There is no easy answer to this. We may have to consider preemptive nonnuclear attacks on nuclear facilities, such as the nuclear fuel reprocessing plant in North Korea. On this I disagree with Senators Robert Byrd and Edward Kennedy, who have called on the United States to respect an absolute ban on preventive attacks. There have been times when preventive war would have been necessary and proper—for instance, in March 1935, when Germany announced that it was tearing up the Versailles Treaty and building a military air force.

It would help if the United States could act against nuclear proliferation with clean hands. Under the terms of the 1970 Nuclear Nonproliferation Treaty, we are committed to deemphasize the role of nuclear weapons and work toward their elimination. But there are signs that the Bush administration is trying to revive the idea that nuclear weapons are for use and not just for deterrence. The administration's Nuclear Posture Review has called for the development of Robust Nuclear Earth Penetrators—nuclear weapons for attacking underground facilities (even though such weapons can't be used without creating severe nuclear fallout)—and the new budget contains a small appropriation for this purpose. The chair of the Defense Science Board has called for a study of nuclear-armed antiballistic missile interceptors. White House Chief of Staff Andrew Card has said that the United States would not rule out the use of nuclear weapons in Iraq. President Bush has announced that he will not seek to ratify the Comprehensive Test Ban Treaty. For a nation with an overwhelming superiority in conventional arms, the development of nuclear weapons for actual use seems counterproductive to the point of insanity.

Some say that nuclear testing is needed to maintain safety and reliability, but both a committee of the National Academy of Sciences in 2002 and the Council of the American Physical Society in 2003 have concluded that it is possible to maintain confidence in the safety and reliability of the existing nuclear weapons stockpile without actually producing nuclear explosions. Indeed, when we tested nuclear weapons in the past, it was usually to develop new weapons.

Personally, I don't think it would be so bad if nuclear weapons on all sides did become somewhat unreliable. We might not then be able to use them for preemptive attacks or bunker busting or missile defense, but what effect would it have on deterrence if there was a possibility that some fraction of our weapons would not achieve the nominal yield? Meanwhile, nuclear proliferation continues: North Korea today, Iran tomorrow. Even in Brazil, a cabinet minister has called for a nuclear weapons program.

You may not realize it, but so far in this talk I have been looking on the bright side. A nuclear attack by terrorists or rogue states could do terrible damage and kill millions of people, but it would not destroy our country. Only one thing could do that: a mistaken attack on our country by the huge Russian arsenal of nuclear weapons.

It may seem terribly "retro" to mention this danger—akin to suggesting that a modern politician would worry about nineteenth-century issues like bimetallism or free love. Granted, in the present state of international relations, no one thinks that either Russia or the United States would ever plan a first strike against the other. Nevertheless, the strategic nuclear forces of both sides remain frozen in their cold war posture. Each is tasked with the responsibility of being able to respond to an attack by the other side before a single attacking nuclear weapon can reach its own land-based missiles and control centers. This means that the decision to attack must be made in minutes, before any nuclear weapons have actually exploded. It takes only two minutes to launch our own land-based intercontinental ballistic missiles, and less than fifteen minutes to launch our submarine-based missiles.

The pressure to decide quickly is more severe for the Russians than it is for us, because they have little left of the invulnerable part of their deterrent (their missile submarines rarely go to sea), and their land-based missiles are vulnerable to a relatively short-range attack by U.S. submarines. In January 1995 the Russian attack decision process was triggered by the launch of a U.S. research rocket from a Norwegian offshore island to study the Northern Lights. The rocket firing was originally mistaken for a launch from an American submarine in the Norwegian Sea, with the separation of multiple stages perhaps giving the impression of an attack by several missiles. The Russian response process was stopped only a few minutes short of their ten-minute deadline for a final decision. (Similar episodes occurred in the Soviet Union in 1983 and in the United States in 1979 and 1980.) The pressure on the Russians for quick decisions will become greater as the United States deploys and improves its antimissile system, which could be thought to have some capability against a ragged Russian second strike.

Defense Secretary Donald Rumsfeld has upheld the deployment of an ineffective missile defense system by saying that it is better than nothing, but in fact it is worse than nothing. Major General Pavel Zolotarev, past deputy chief of staff of the Russian Defense Council, has said that U.S. missile defense plans make it harder for Russian nuclear planners to consider deep cuts in their arsenal coupled with dealerting. Can we really assume that Russian judgments about whether they are under attack will always be made correctly, especially if relations between the United States and Russia sour in the future?

Several steps have been taken to ameliorate this danger, all sharing the common feature of being ineffective. In May 1994 Presidents Clinton and Yeltsin agreed that the United States and Russia would stop targeting each other's territory. This is a bad joke; the targeting can be restored in seconds. In 1998 the presidents of the United States and Russia agreed to establish a center in Moscow for the exchange of data on rocket launches. Plans for this center were completed, but it was never brought into operation. In March 2003 the Senate ratified the Strategic Offensive Reductions Treaty, which had been signed the previous year by Presidents Bush and Putin. It requires a reduction in the number of strategically deployed nuclear weapons on both sides, but the treaty will reduce the numbers only to about two thousand weapons on each side by 2012, and the delivery vehicles and thousands of weapons taken out of service will not need to be destroyed, only separated. We need to reduce the number of nuclear weapons on both sides to hundreds, not thousands; to count all weapons, not just those that are strategically deployed; and to take these weapons off hair-trigger alert. Nothing is more important. In any one year, the danger of nuclear attack by mistake is small, and aside from the warnings issued by a few hardy souls (such as Bruce Blair, the director of the Center for Defense Information, and former senator Sam Nunn), it receives little attention. No president of either party has given this danger a high priority. But it is always with us, and in the end it may destroy us.

12

What Price Glory?

I read history for pleasure, and a good deal of what I read is military history. It seems to me that in the history of war, as in the history of science, it is possible to reach pretty clear judgments of success or failure that are usually impossible in social or political history. For this reason, military history can provide especially useful insights into the strengths and weaknesses of societies engaged in war. That, at any rate, is the reason I usually give for what might otherwise seem like a morbid preoccupation for a peace-loving professor. Perhaps another reason for my interest in military history is that I was a boy during World War II, bitterly regretting that I was too young to join the fight.

For some time my reading in recent years had given me the idea that choices of military tactics and technology were often made to enhance the glory of military commanders, even though in reality they did not increase the prospects for avoiding or winning wars. I could see examples of this also in current military planning, some of which had concerned me professionally or politically. I had the idea of writing an article about the part that glory-seeking had played in military history, but felt some hesitation about it, as I have no professional credentials as a historian. Then Robert Silvers invited me to submit an article on any subject as a contribution to the 40th anniversary issue of the *New York Review of Books*, to be published in November 2003, and I took the opportunity of expressing my view of military glory in the following article. Books cited in this article are listed at the end.

War offers ample opportunities for most varieties of foolishness. Among these, there is one sort of folly to which war is especially well suited: the lust for glory. One can hardly ever be sure about a commander's motives in any one case, but there are familiar signs of that lust: a readiness to accept a challenge to fight under unfavorable circumstances; a preference for taking action independent of allies or colleagues; an unreasoning predisposition for offense rather than defense; and an effort to seize a decisive role for one's self in winning victory. Examples come easily to mind. Antony accepted Agrippa's challenge to fight by sea at Actium, though he was stronger by land. In 1421 the Duke of Clarence violated the orders of his brother, King Henry V, and died attacking five thousand French troops with 150 mounted men-at-arms and no archers. To recapture the glory he had won by riding around McClellan's army in search of its flank during the defense of Richmond in 1862, J. E. B. Stuart in June and July of 1863 led his cavalry on a wild ride through Maryland and Pennsylvania, even though it left the Army of Northern Virginia without the reconnaissance it needed in the week before the Battle of Gettysburg. Admiral William F. Halsey Jr. commanded the Third Fleet to chase Japanese battleships and carriers while other Japanese battleships threatened American soldiers landing on the beaches of Leyte Island.

Though there always will be soldiers and sailors "seeking the bubble reputation, even in the cannon's mouth," it seems that the vainglory of individual commanders has lately become less dangerous in war, as improvements in the technology of communications and surveillance have increased the ability of senior commanders to control subordinate officers. But there is a continuing danger from an *institutionalized* vainglory. Sometimes a branch of the military may try to maximize its opportunity for glory, turning its back on other less glamorous tasks that are really needed. This can become an ideology, like the French army's doctrine in 1914 of *"l'attaque à outrance."* The military may even adopt weapons that serve more to enhance its glory than the likelihood of victory, and weapons themselves may become imbued with a glamour that stands in the way of sensible decisions about their use. One can find instances throughout history, and they extend unfortunately to the present day, with dangerous effects on our current defense policy.

On February 1, 1917, Germany began a program of unrestricted submarine warfare. The effect on British shipping was devastating. During the first three months German U-boats sank 844 ships, at a cost of only ten of their submarines. According to Winston Churchill, "That was, in my opinion, the gravest peril that we faced in all the ups and downs of that war."

It should have been obvious that the solution to the U-boat threat was to require merchant ships to sail in convoy. As Churchill later explained in *The World Crisis*,

> The size of the sea is so vast that the difference between the size of a convoy and the size of a single ship shrinks in comparison almost to insignificance. There was in fact nearly as good a chance of a convoy of forty ships in close order slipping unperceived between the patrolling U-boats as there was for a single ship; and each time this happened, forty ships escaped instead of one.

(This is also the reason that fish of many species swim in schools.) Furthermore, forty merchant ships can be guarded by a much smaller number of destroyers or other escorts, while it would be impossible to send an escort with each merchant ship sailing alone. Even before sonar became available, a submerged submarine could be found when it attacked, by tracing back the track of the submarine's torpedo. It is true that a convoy presents a great many more targets than a single ship, but even with all those targets, a single U-boat can destroy only a few ships before it exhausts its torpedoes or is destroyed or driven off by the escorts. U-boats in World War II learned to call in other U-boats to the attack when a convoy had been found by using radio to communicate with their headquarters in occupied France, but this technology was not available in World War I. (Nor is it available to fish.)

For several months after the start of unrestricted submarine warfare, while British ship losses mounted, the Admiralty continued to reject the use of convoys. For the Royal Navy in World War I, convoy duty was inglorious.[1] Arthur Marder, the leading historian of naval warfare in the early twentieth century, has explained in *From the Dreadnought to Scapa Flow* that

> the strange dogma had emerged in the pre-war generation that to provide warship escorts to merchant ships was to act essentially "defensively" (because it protected ships from attack), which was *ipso facto* bad, and that to use naval forces to patrol trade routes, however futile the result, was to act "offensively" against the warships of an enemy, and this was good.

In 1915 British Vice Admiral H. F. Oliver, chief of the War Staff, had explained to the secretary of the War Cabinet, Sir Maurice Hankey, that the correct antidote to the U-boat menace was not convoy, but "hunting."

The trouble is that in the absence of long-range aviation and radar (and often even with them) it is very difficult for patrols to find submarines cruising in the open ocean. In the days when I used to work on antisubmarine warfare as a member of JASON, a group of academics serving as defense consultants, my efforts were mostly limited to solving equations, but in the early 1960s I did once go out on antisubmarine maneuvers. I was on a destroyer escort searching for a World War II diesel submarine that had left Key West that morning. Our ship and a destroyer and aircraft searched all day, using passive and active sonar and magnetic detection devices, but it could not find the sub. At the end of the day the submarine had to send up a radio beacon to tell us where it was. Remembering this, when I later read the estimate of Rear Admiral William Sims (the American admiral in charge of U.S. naval forces in Europe during World War I) that for patrols to find sub-

---

1. To be fair, more sensible arguments were offered against convoys. It was said that merchant ships could not sail in tight formations, but when they tried it turned out that they could. It was said that there were not enough convoy escorts, but it turned out that the destroyers available could do the job.

marines, which can submerge at will, requires about one destroyer per square mile of ocean, I thought that Sims had been too optimistic. And of course there are a lot of square miles in the ocean. According to Sims's estimate, to find a submarine cruising somewhere in the western approaches to the British Isles would have required twenty-five thousand destroyers.

As aggressive patrolling clearly failed to counter the U-boats and shipping losses grew, the Admiralty became desperate. When Sims arrived in London on April 10, 1917, he was invited to meet with the First Sea Lord, Admiral Sir John Jellicoe. Jellicoe told him that if shipping losses continued at their present rate Britain would have to leave the war, and that the Admiralty had no idea what to do about it. Sims cabled back to Washington that "briefly stated, I consider that at the present moment we are losing the war."

Pressure from Sims, Hankey, and Lloyd George finally forced the Admiralty to try using convoys. As an experiment, a convoy of merchant ships was sent from Gibraltar to Britain, and all ships arrived safely on May 20, 1917. The next day the Admiralty at last decreed that all merchant shipping to or from Britain must travel in convoys, and the rate of shipping losses dropped sharply.

In World War II it was the strategic air forces of Britain and the United States that tried to play a glorious but unrealistic role—to win the war by themselves through strategic bombing, i.e., bombing aimed at the enemy's industry and population. There has been a great debate about the efficacy of strategic bombing in World War II. But however useful strategic bombing may have been, it could not have won the war by itself, at least not until the advent of nuclear weapons. In the invasion of Normandy, it was surely necessary that all available bombers and their fighter escorts should come under Eisenhower's authority, so that he could divert them as needed from strategic bombing to support of ground troops and the interdiction of German reinforcements. Nevertheless, James Doolittle and Arthur Harris, the commanders of the American Eighth Air Force and the British Bomber Command, put up a strong resistance to this arrangement, or to any pause in strategic bombing, and they managed to enlist Churchill's support.[2] To gain control over the strategic air forces during the critical phase of the invasion, Eisenhower had to state that as long as he was in command he would accept no other arrangement, thus implicitly threatening to resign.

Even Stuart's failure to keep in touch with Lee before Gettysburg may have represented an institutionalized hubris rather than mere personal vainglory. Other Civil War cavalry generals also liked independent action. In August 1864 the commander of the Confederate Army of Tennessee, John B. Hood, sent his cavalry corps, under General Joseph Wheeler, to raid the railroad lines that were supplying Sherman's army in Georgia, with the expectation (according to Richard McMurry's book *Atlanta 1864*) that Wheeler would be gone for only a few days. As it happened, nobody in the Army of Tennessee would see Wheeler's troops again until mid-October. And in February 1864 the Union cavalry general Judson Kilpatrick talked Lincoln into allowing his cavalry to make a disastrous raid into Richmond in order to distribute pamphlets and free some prisoners. As Bruce Catton observed, "It was born of a romantic dream and it was aimed at glory, and glory was out of date."

We have recently had another example of the military distaste for unheroic roles. Just as the Royal Navy preferred "hunting" to convoy duty in World War I, and the Allied air forces preferred strategic bombing to ground support in World War II, and the cavalry preferred independent action to the support of infantry in the Civil War, so in recent years the United States Army has preferred

---

2. Eisenhower in *Crusade in Europe* says that it was Doolittle, Harris, and Churchill who opposed giving him authority over the strategic air forces during the invasion, while Gerhard Weinberg's history of World War II, *A World at Arms*, attributes the opposition also to Ira Eaker, the commander of the Fifteenth Air Force in Italy, as well as to Doolittle and Harris. Russell Weigley in *Eisenhower's Lieutenants* cites only Harris as resisting Eisenhower's authority. If Weigley is right, then British national pride as well as air force vainglory may have played a role in this dispute.

to plan for fighting battles without worrying about how to govern conquered territory. The Army in World War II had an effective Division of Military Government. It was established in the Office of the Provost Marshall in July 1942, long before there were any captured Axis territories to govern. It was this division and the personnel whom it trained at the Charlottesville School of Military Government that made it possible for the United States later to govern Japan and parts of Germany and Italy in an orderly way, without encountering widespread looting, rioting, or guerrilla attacks.

In the years after the war, responsibility for military government was relocated in the Civil Affairs branch of the Army. Support for this branch was allowed to dwindle, and Civil Affairs survived several attempts to disband it as a separate unit, until in 1987 it finally found a home in the Special Operations Command. There it had to fight off attempts to divert its remaining funds and personnel slots to Special Forces. At the end of the 1980s, an Army-commissioned report, in a chapter called "Pruning Non-Essentials," asked the questions "Should 7,000 reservists continue to be trained to govern occupied nations? Is there a need for those trained in the administration of art, archives, and monuments to preserve the culture of occupied territories?"[3] Civil Affairs became known as a dead end for career officers.

There is now just one active-duty Civil Affairs unit, the 96th Civil Affairs Battalion (Airborne), headquartered at Fort Bragg; the remaining 95 percent of Civil Affairs personnel are reservists. In Afghanistan there are now only about two hundred Civil Affairs personnel, as compared with about 15,000 military government soldiers in the American Zone of Germany soon after the German surrender in World War II. A colonel (not in Civil Affairs) who is just back from Iraq tells me that there are about two thou-

---

3. Quoted by Stanley Sandler, *Glad to See Them Come and Sorry to See Them Go: A History of US Army Civil Affairs and Military Government, 1775–1991* (USASOC, 1998), 372.

sand Civil Affairs officers there (not all in military government), leaving few anywhere else, and that although they are doing good work, there are not nearly enough of them. Unfortunately the Defense Department's priorities do not seem to have changed. Later this year it plans to close the ten-year-old Peacekeeping Institute of the Army War College.

Histories of military technology often describe how the adoption of newly developed weapons is delayed by cultural influences. (Perhaps the most striking example was the refusal of the Tokugawa shoguns to allow guns in Japan.) But in these histories, when new weapons are finally adopted, it is always supposed to be because of their objective effectiveness, rather than something as irrational as a lust for glory.

In his influential 1962 book *Medieval Technology and Social Change,* the distinguished medieval historian Lynn White follows this pattern. As an example of new military technology, he chose the combination of stirrup and lance. The foot-stirrup was unknown in classical antiquity. As White tells the story, it was invented in China sometime around the fifth century AD. Nomadic horsemen brought the stirrup through Central Asia to Constantinople. Nowhere along this journey did it have much effect on how battles were fought. But when the stirrup reached Western Europe late in the seventh century its effect was explosive. The stirrup allowed a horseman with "couched" lance, i.e., one held between the body and the upper arm, to deliver the whole force of a charging horse at the tip of his lance, without losing his seat. White argued that suddenly it was necessary to recruit horsemen who could afford the armor and war horses needed to fight in this way, horsemen who could only be paid by the grant of lands. Starting in 732, Charles Martel, the great mayor of the palace of Merovingian France, began to seize church lands and grant them to his cavalry. A new class of feudal knights and nobles was born.[4]

---

4. Michael Howard endorses this account in *War in European History.*

I have my doubts about this story. There is no question that stirrups are a good thing, as everyone knows who has ever tried to get on a horse. I don't doubt that the cavalry charge with couched lance became a popular tactic after 1100. But was it adopted because it was effective?

The use of the stirrup that White thought was "above all" important in tipping the balance in favor of cavalry over infantry, enabling a mounted knight to charge with couched lance, which is also the one tactic in which the stirrup is indispensable, does not seem to have been very effective against infantry. In such a charge each knight can only kill one foot soldier before a general melee begins in which the lance is useless. A "destrier," a horse strong enough to carry an armored knight and bred and trained to charge an enemy, was enormously expensive. As the Chorus says in *King Henry the Fifth*, "they sell the pasture now to buy the horse." When we consider also the cost of the knight's and horse's armor, and of the necessary grooms and remounts, killing a foot soldier this way seems to have been hardly cost-effective. It would be like tanks without guns attacking infantry by running them down, one by one. Doubtless the tanks would defeat an equal number of foot soldiers, but in a war between two countries with equal resources, the side that used its resources this way would surely lose.

Also, foot soldiers were not defenseless in the Middle Ages. They could stand behind pointed stakes set in the ground, as the English archers did at Agincourt and other battles of the Hundred Years' War. They could protect themselves with long pikes, as Italian foot soldiers did to repulse the cavalry of Emperor Frederick I at Legnano in 1176, or fight on marshy ground, as Flemish pikemen did in repulsing the French cavalry at Courtrai in 1302, or on hilly ground, as the Swiss did in defeating Leopold of Austria at Morgarten in 1315. And the foot soldiers could fight back with missile weapons. A bolt from a crossbow or an arrow from a long bow could penetrate all but the best armor. It was English archers who defeated the French knights at Crécy, Poitiers, and Agincourt.

I do not know of a single European battle in the Middle Ages that was won by a charge of cavalry with couched lances against a line of foot soldiers. No clear examples are mentioned by Charles Oman in his classic two-volume *A History of the Art of War in the Middle Ages*. Granted, foot soldiers all but disappeared from European battlefields in the high Middle Ages, from 1100 to 1300, but apparently not because they had been defeated by the lance and stirrup.[5]

Oddly, White mentions only one European battle to prove the superiority of the new military technology, the Battle of Hastings on October 14, 1066. It is true that the Normans under Duke William were mostly mounted at Hastings, and they defeated King Harold's English army, which fought on foot. But the English soldiers were probably outnumbered by the Normans, and they were exhausted. After defeating an invasion by a Norwegian army under Harald Hardraade at Stamford Bridge on September 25, they had arrived in London on October 6, having marched the two hundred miles from York, and they then marched another fifty-eight miles to the neighborhood of Hastings. What is more, it is not clear that the Norman cavalry at Hastings ever charged English foot soldiers with couched lance.[6] The only reliable testimony to the weapons used at Hastings is the Bayeux tapestry, embroi-

---

5. Oman mentions two battles that encouraged the nobility after the revival of infantry in the fourteenth century to believe in the continued superiority of cavalry over infantry: Mons-en-Pevele in 1304 and Cassel in 1328. As pointed out recently by Kelly DeVries in *Infantry Warfare in the Early Fourteenth Century*, though French cavalry did defeat Flemish infantry in both battles, this was only after the Flemings had become disorganized by ill-conceived charges against the French. At Mons-en-Pevele an initial French cavalry charge was turned back by the Flemish infantry, and at Cassel the French did not even try to attack the Flemish line until after the Flemish attack.

6. White does not say in so many words that the Normans charged the English with couched lances at Hastings, but he refers to Hastings as the "most spectacular" application of the new military technology, and as "a conflict between the military methods of the seventh century and those of the eleventh century."

dered not long after the battle.[7] Looking over reproductions of the whole tapestry, I see plenty of lances in the hands of Norman cavalry, but that proves nothing. Lances had been widely used by cavalry for stabbing and throwing since long before the invention either of stirrups or feudalism. Many of the lances shown in the Bayeux tapestry are held above the head, as if they are about to be thrown.

Just one panel, number 65, shows what may be a couched lance being used to attack an English foot soldier, but there is no way to tell if this is the result of a cavalry charge. I find it telling that none of the lances shown on the Bayeux tapestry have the kind of handhold (such as the conical handholds known as vamplates that are shown on later illustrations of tournaments) that would allow a knight on a charging horse to drive the lance into an enemy. Without such a handle, a lance striking anything would just slide through the knight's armpit, stirrups or no stirrups. In any case, the English were defeated at Hastings more by Norman archers than by knights. The Bayeux tapestry shows an Englishman, generally believed to be Harold, struck in the eye by an arrow.

So if the cavalry charge with couched lance was an ineffective tactic against infantry, why did the lance become the standard weapon of medieval cavalry, and why did cavalry dominate the battlefields of the high Middle Ages? It seems likely to me that, instead of the new class of feudal nobility being called into being to take advantage of the combination of stirrup and couched lance,

7. There are no eyewitness written accounts of the Battle of Hastings. The most frequently quoted contemporary accounts are *The Deeds of William*, by the Duke's chaplain, William of Poitiers, and *The Song of the Battle of Hastings*, attributed to Guy, Bishop of Amiens. They mostly mention arrows, javelins, and swords. There are a few places where English translations refer to lances (none of them appearing at the same point of the battle in both accounts), but it is not always clear that this is what was meant in the Latin originals. The Latin word *hasta* can mean either lance or spear. Also, nowhere in either account is it clear that an attack was made with the lance couched. The translators of William of Poitiers comment that the use of the couched lance was restricted in this battle, because of the hilly terrain.

feudalism arose for other reasons,[8] and the cavalry charge with couched lance became the favorite tactic of the feudal knights and nobles because it gratified their desire for military glory. Landowners could see themselves as military heroes, not merely as recruiting sergeants of the prince. The great expense of this tactic was an advantage, not a drawback, because peasants could not afford a destrier and armor. The reason that foot soldiers were not used in battle is not that they would have been ineffective, but that they were common. If this all sounds too illogical to be true, recall Oman's verdict on the knights of the high Middle Ages: "A feudal force presented an assemblage of unsoldierlike qualities such as have seldom been known to coexist."

The cavalry charge with couched lance found its real use in tournaments. These gorgeous rituals gave skillful knights like England's William Marshall the opportunity to become international celebrities. It is sometimes said that tournaments became popular in the twelfth century as training grounds for real battle,[9] but it may be equally true to say that cavalry battles with infantry excluded served as training grounds for tournaments.

It is not hard to find examples closer to the present of weapons that were adopted more for glory than for military effectiveness. Between the world wars, officers of the U.S. Army Air Corps imagined that land-based aircraft would not only be able to destroy the enemy's industry by strategic bombing but would also be able to conquer enemy fleets. Air Corps General William Mitchell predicted that air power "will not only dominate the land but the sea as well." Just as medieval knights were supposed to make ple-

---

8. For instance, according to the much-debated thesis of Henri Pirenne, it was the closing of the Mediterranean to European trade by the Arabs that was responsible for the replacement of Roman civilization in the West with feudalism.

9. Wellington is supposed to have remarked that the Battle of Waterloo was won on the playing fields of Eton. But, being smarter than medieval generals, he knew better than to arm his soldiers with cricket bats.

beian foot soldiers unnecessary, bombers would make navies irrelevant.

For bombers to act independently at sea beyond coastal waters without having to be carried to the attack by naval vessels, they would have to have a very long range and carry a large bomb load, just as they would for strategic bombing. Such an airplane would have to be large, so it could not be used as a dive bomber, and it would therefore have to release its bombs in level flight. To avoid antiaircraft fire it would have to bomb from a high altitude. After a competition among several aircraft manufacturers, the Boeing corporation in January 1936 received a contract for a bomber to meet these requirements. It would eventually be named the B17, or "Flying Fortress," and after the war it would become the glamorous star of the movies *Memphis Belle* and *Twelve O'Clock High.*

The B17 met all technical specifications of the Department of the Army, but it could not fulfill the purposes for which it was planned. After the attack on Pearl Harbor, B17s were sent to Midway Island. As a Japanese fleet approached Midway in June 1942, it was spotted by B17s. In fifty-five sorties between June 3 and 5, the B17s dropped 315 bombs, but despite optimistic reports from the pilots, not a single one of these bombs struck a Japanese ship. Antiaircraft fire from the Japanese fleet kept the B17s above twenty thousand feet, and at that altitude it was just too hard to drop a bomb on a target as small as a ship. Also, like medieval foot soldiers dodging an attack by knights with couched lances, ships can maneuver out of the way of bombs dropped from high altitude. As Samuel Eliot Morison remarked, in his *History of United States Naval Operations in World War II,* the B17s would have done more damage by tracking the Japanese ships than by trying to bomb them, but their pilots were trained only for bombing.

The confidence of the Army Air Force in its ability to bomb enemy ships arose in part from a dramatic demonstration staged in 1921 by Mitchell, in which Army bombers had successfully sunk a number of surrendered German warships, including the dreadnought *Ostfriesland*. But the German ships were at anchor, not maneuvering, and since they were not firing at the aircraft, it was safe to bomb them from low altitude. The four Japanese carriers at Midway were sunk not by B17s but by short-range Navy SBD dive bombers, which had been carried to the battle by *Enterprise* and *Yorktown*. During 1943, B17s were mostly withdrawn from the Pacific theater, though a few were kept for less showy activities like dropping life rafts to shipwrecked sailors.

The B17 also encountered trouble even in its primary role of strategic bombing. To have any chance of hitting targets like factories and railroad marshaling yards without our present satellite-based global positioning system, bombers had to attack in daylight. But flying over enemy territory in daylight made them vulnerable to attack by enemy interceptor airplanes as well as antiaircraft artillery. For this reason the design of the B17 had sacrificed part of its bomb capacity to enable it to carry exceptionally heavy armament: thirteen machine guns, manned by six crew members. (The maximum bomb load of the B17 was less than that of the British Lancaster[10] or the American B24. The B24 proved more useful than the B17, but it never attracted the same public attention.) The Air Force expected that B17s flying in tight formation would be able to fight off interceptors by themselves and reach targets beyond the range of American fighter escorts. Experience did not bear this out. In a daylight raid on August 17, 1943, a force of 315 unescorted B17s attacked a ball-bearing factory at Schweinfurt and a Messerschmitt factory at Regensburg, and lost sixty bombers and their precious crews. At that rate of loss, the Eighth Air Force could not have continued daylight strategic bombing for more than a few more raids without losing almost all

---

10. The Royal Air Force had been even more enthusiastic about strategic bombing than the U.S. Army Air Force, but they were more realistic about the ability of bombers to defend themselves. They gave up the goal of precision bombing and attacked German cities at night, hoping to weaken Germany by damaging morale and workers' housing.

of its strength. Daylight strategic bombing deep into Germany had to be suspended, until long-range escort fighters like the P51 became available in February–March 1944.

Despite all disappointments about the ineffectiveness of the B17 in performing the tasks for which it had been planned, it embodied the dreams of glory of the Air Force, and became a symbol of American air power. Over 12,000 B17s were built during the war. It is true that B17s, along with other bombers, did take an important part in the war by attacking German synthetic oil plants and transportation facilities after long-range escorts became available, and above all in forcing Germany to divert artillery and fighter aircraft from battlefields to air defense. But the huge commitment of American resources to the B17 limited other production. Allied war planning was continually constrained by a shortage of landing craft, but never by a shortage of B17s.

Since World War II it has been rockets and nuclear weapons that have the glamour that used to surround bombers or lances. There is no doubt about the extraordinary effectiveness of modern high-technology weapons in certain circumstances. But it always needs to be asked at which tasks are they effective, and which of these tasks actually need to be accomplished?

In the decades since nuclear weapons were used in war, there has developed a healthy conviction that these weapons should not be used again for anything but deterrence. In a remarkable international bargain, a number of countries agreed in 1970 that they would not develop nuclear weapons at all, in exchange for a pledge by the United States and other nuclear powers to deemphasize the role of nuclear weapons and to work toward their elimination. Now the Bush administration has turned its back on the Non-Proliferation Treaty by calling for work on a new generation of low-yield nuclear weapons. This would add nothing to our nuclear deterrent, but might actually be used in fighting wars. Prominent in the administration's wish list is a "robust nuclear earth penetrator," a nuclear weapon that could penetrate into the earth to attack subterranean bunkers and laboratories.

As with the couched lance, it is hard to think of a plausible mission in which new low-yield nuclear weapons would actually be effective. As I have discussed previously,[11] a nuclear weapon carried underground by a robust earth penetrator would only be able to attack targets that are not buried too deeply and whose location is precisely known, and any such attack would inevitably produce large quantities of radioactive fallout at the surface, possibly killing large numbers of civilians.

The problem is not so much that the money spent on new nuclear weapons would be wasted. The cost of these programs is in the range of tens or hundreds of millions rather than billions of dollars. The problem is, rather, that the United States would be violating the 1970 Non-Proliferation Treaty and encouraging a new round of nuclear weapons development throughout the world. For a nation with an enormous lead over the rest of the world in non-nuclear military technology, this would be foolishness on a scale that even medieval knights might find implausible.

National missile defense has for some the allure of a good science fiction movie. A fleet of American satellites detects the launch of a missile by some evil regime in Asia. Powerful radars in Alaska track the missile's warhead, and American missiles are launched to meet it above the Earth's atmosphere. Our missiles then release exoatmospheric kill vehicles (EKVs), which guide themselves to collide with the enemy warhead, and destroy the bomb it contains. Seattle is saved.

Unfortunately, missile defense is not likely to work so neatly.[12] So far, tests of the EKV have alternated between successes and failures, and now the tests have been suspended while work continues on the booster that would carry the EKVs into space. But even if the tests had an unbroken series of successes, they would still have

11. "The Growing Nuclear Danger," *New York Review of Books*, July 18, 2002; chapter 8 in this collection.
12. I discussed this in more detail in "On Missile Defense," *New York Review of Books*, February 14, 2002; chapter 7 in this collection.

the same quality of phoniness as the destruction of the *Ostfriesland* by bombers in 1921. It is not that tests like this have no value, but rather that they can't be relied on to tell us whether the system being tested would be really effective. Just as the *Ostfriesland* in Mitchell's test was neither maneuvering nor shooting at the bombers attacking it, none of the warheads targeted in the tests of the EKV has been accompanied by the simple realistic decoys that could defeat the EKVs. Nor have other possible countermeasures against EKVs been addressed in any of the tests so far. In the real world, B17s could not hit Japanese warships that were maneuvering in unpredictable ways, and similarly our EKVs will not be able to destroy incoming warheads if the enemy uses countermeasures that we cannot know in advance.

The proponents of missile defense to whom I have talked generally acknowledge these problems, but they argue that we must at least make a start, and then learn as we go along. As Secretary Rumsfeld has put it, the proposed national missile defense system is at least better than nothing. I think a strong case can be made that it is worse than nothing. It will hurt relations with our allies, discourage Russia from taking its missiles off ready alert, and encourage China to increase its missile forces more than had been planned.

A "boost-phase intercept" system, which relies on attacking missile boosters soon after launch, would not be subject to these criticisms if it defended our allies as well as ourselves, and if it were located so that it could not reach missiles launched from Russia or China. But a recent study by a panel of the American Physical Society indicates that feasible interceptors would not be able to reach missiles launched from North Korea or Iran before warheads separate from the booster[13]

Granted, one cannot be sure about how other countries will respond to an American missile defense system. But there is no ques-

13. Report of the American Physical Society Study Group on Boost-Phase Intercept Systems for National Missile Defense," July 2003, available at www.aps.org.

tion about the enormous cost of missile defense. We are currently spending about nine billion dollars a year just for research and development, and a deployed system covering the entire United States would surely cost several hundred billion dollars, all to ineffectively counter a highly implausible threat. Any missile attack on the United States would immediately reveal which country launched it and expose it to devastating retaliation. Such deterrence protected the United States from nuclear attack for almost half a century of cold war. It is true that without missile defense the United States might be deterred from trying to overthrow a "rogue state" that possessed nuclear-armed intercontinental missiles. But for the United States to take such an action without fear of nuclear retaliation, a missile defense system would have to be nearly perfect, not merely better than nothing.

Even those for whom national defense is the one clearly legitimate reason for government spending ought to consider whether the enormous sums required for missile defense would not be better spent on defense of other sorts. There are many ways to attack the United States with nuclear or biological weapons that, unlike ballistic missiles, do not immediately reveal the source of the attack. Over the past year or so I have served on two panels of the Council on Foreign Relations, the Hart-Rudman Independent Task Force on Homeland Security Imperatives and the Rudman Independent Task Force on Emergency Responders. It has been painful to learn how much the lack of funds has limited our ability to defend the country from terrorists. Physical security at our seaports is grossly underfunded. Also, the United States may be spending one third of what is required to adequately provide for those who would have to respond to emergencies. American cities have fewer policemen and firemen now than they did before September 11, 2001.

We face another danger, even greater than that from terrorists. For all the good relations between Russia and the United States today, Russian nuclear forces are frozen in a cold-war configuration, one designed to respond to warning of an American attack within

ten minutes by a massive nuclear counterattack, before a single nuclear weapon can reach Russia's land-based missiles or control
centers. This puts not just a city or two but the entire United States
in danger of irreversible destruction by mistake, a danger that will
increase as Russia's capacities to detect an attack become more
and more degraded. Much could be done to lessen this danger,
such as seriously reducing nuclear arsenals on both sides and sharing information about missile launches, but this has not been a priority of any U.S. administration. Reducing the threat of nuclear
attack by mistake, improving port security, providing for emergency
responses, bringing effective government to Iraq and Afghanistan—
in none of these do our leaders find sufficient glory.

AMONG THE BOOKS DRAWN ON FOR THIS ESSAY

A Stillness at Appomattox: The Army of the Potomac, vol. 3, by Bruce Catton
    (Anchor).
The World Crisis, vol. 4, by Winston S. Churchill (Scribner).
Infantry Warfare in the Early Fourteenth Century, by Kelly DeVries (Boydell
    and Brewer, distributed in the US by University of Rochester Press).
Crusade in Europe, by Dwight D. Eisenhower (Johns Hopkins University Press).
The Carmen de Hastingae Proelio of Guy, Bishop of Amiens, translated and ed-
    ited by Catherine Morton and Hope Muntz (Clarendon Press).
War in European History, by Michael Howard (Oxford University Press).
From the Dreadnought to Scapa Flow, vol. 4, by Arthur J. Marder (Oxford Uni-
    versity Press).
Atlanta 1864: Last Chance for the Confederacy, by Richard M. McMurry (Uni-
    versity of Nebraska Press).
*Winged Defense: The Development and Possibilities of Modern Air Power—
    Economic and Military,* by William Mitchell (Dover).
Coral Sea, Midway, and Submarine Actions, May 1942–August 1942, by
    Samuel Eliot Morison (Little Brown; part of a fifteen-volume set).
A History of the Art of War in the Middle Ages, by C. W. C. Oman (Burt
    Franklin, two volumes).
The Art of War in the Middle Ages, AD 378–1515, by C. W. C. Oman, revised
    and edited by John H. Beeler (Cornell University Press).
Mohammed and Charlemagne, by Henri Pirenne (Dover).
Hankey: Man of Secrets, Vol. 1, 1877–1918, by Stephen Roskill (Naval Insti-
    tute Press).

The Victory at Sea, by William S. Sims (James Stevenson).
The Bayeux Tapestry: A Comprehensive Survey, edited by Frank Stenton (Phaidon).
Eisenhower's Lieutenants: The Campaign of France and Germany, 1944–1945, by Russell F. Weigley (Indiana University Press).
A World at Arms: A Global History of World War II, by Gerhard L. Weinberg (Cambridge University Press).
Medieval Technology and Social Change, by Lynn White (Oxford University Press).
The Gesta Guillelmi of William of Poitiers, translated and edited by R. H. C. Davis and Marjorie Chibnall (Oxford University Press).

The *New York Review of Books* published two interesting letters commenting on the foregoing article. One letter, by Professor R. Howard Bloch of Yale's French Department, was published in December 2003. Professor Bloch disagreed with my statement that the cavalry charge with couched lance was not an effective military technique in the Middle Ages. He pointed out that the Normans who fought at the Battle of Hastings thought highly enough of this technique to bring horses with them across the Channel. He further claimed that Normans had used the couched lance in Southern Italy before the Conquest, most dramatically in their defeat of a papal army at Civitate in 1053.

The other letter, by Harry Lieber, was published the following April. Mr. Lieber acknowledged that I might be right about the ineffectiveness of the cavalry charge with couched lance in the Middle Ages, but argued that this technique won the day at the Battle of Vienna in 1683. He pointed out that the Polish heavy cavalry under King Sobieski broke the Turkish line and sent the Turkish army, largely infantry, into flight. I answered the two letters with the following remarks.

Though I am grateful to Professor Bloch for his compliment about my article "What Price Glory?" in the November 6 issue, I can't agree that I was wrong about the Battle of Hastings. I didn't argue one way or the other about whether Duke William thought that cavalry attacks with couched lances would be effective against infantry. It wouldn't surprise me if he did. There can't have been many military commanders or organizations in history who decided to adopt tactics or weapons that they *knew* to be ineffective because they thought it would maximize their glory. My point, which perhaps I should have made more explicitly, is that the lust for glory can distort judgments about what tactics or weapons *are* effective. My article offered evidence that the cavalry charge with couched lance did not play a significant role at Hastings, perhaps because the hilly terrain would have made this tactic ineffective. I discussed Hastings not because it gives clear evidence one way or the other about the effectiveness of the combination of lance and stirrup, but because this was the one battle that had been mentioned as an example of its effectiveness in Lynn White's *Medieval Technology and Social Change*.

The remark by Professor Bloch about the Battle of Civitate raises a more serious challenge. Did this battle (unlike Hastings) really provide an example of the effectiveness of cavalry charges with couched lances? It is generally agreed that at Civitate a Norman army consisting chiefly of about three thousand cavalrymen defeated a larger papal army composed mostly of infantry. But there doesn't seem to be any evidence that the Normans charged with couched lances, the one tactic for which the stirrup is indispensable. The most frequently cited primary source on the Battle of Civitate is the *Deeds of Robert Guiscard* by William of Apulia. Professor Bloch may have been misled by a 1961 French translation of this work, which tells how the Norman paladin Robert Guiscard pierced through a papal soldier with his lance. I am told by a faculty colleague in the Department of Classics at the University of Texas that the original medieval Latin text only says that the soldier was pierced with a sharp point *(cuspide)*, not necessarily the point of a lance. Later, William describes Robert as fighting with a lance in one hand and a sword in the other, which doesn't suggest a charge with couched lance. Also, whatever tactics were used by the Normans, their victory proves little. The papal army was a rabble, described by Gibbon as a "vile and promiscuous multitude . . . who fought without discipline and fled without shame." It is true that the papal army had a hard core consisting of seven hundred well-trained Swabian foot soldiers, but they withstood one Norman cavalry charge after another until they were overwhelmed by superior numbers. A tactic by which three thousand cavalry can only with difficulty defeat seven hundred infantry can hardly be said to be effective, let alone cost-effective.

I will take this opportunity to update one point in my article. Since the time the article was written, the U.S. Army has reversed its decision to close its Peacekeeping Institute.

Mr. Lieber's letter is extremely interesting. I had not known that in the 1683 battle of Vienna, Jan Sobieski's cavalry charge against the Turkish infantry was with couched lances. I had argued in my article for the ineffectiveness of this tactic in medieval warfare, but this episode serves as a reminder of one way that a cavalry charge with or without couched lance might win a battle—by terrifying ill-prepared foot soldiers out of their wits rather than by doing them any actual harm. That seems to have been what happened at Vienna. The Turkish infantry had been thrown into confusion by an effective charge by Marigny's infantry just before the charge by Sobieski's cavalry. The cavalry, rushing toward them with couched lances, threw the disorganized Turkish infantry into a panic from which they could not be rallied. Similar episodes may have occurred in the Middle Ages. On the other hand, the cavalry charge with couched lance was quite ineffective against well-prepared infantry, like the English archers of the Hundred Years War.

Since this article appeared, events in Iraq and Afghanistan have provided further evidence of the unwisdom of draining resources from the U.S. Army's Civil Affairs division. But as this article makes clear, this was not just the fault of the Bush administration; it is also the result of decisions made by earlier administrations, and by the Army itself.

13

Four Golden Lessons

It is customary in speaking at college commencement exercises to offer advice about life and work to the graduating students. I have generally avoided this practice when I have given graduation talks, because I don't know how to advise students who are heading in many different directions—business, the military, law, scholarship, politics, whatever. It was different, though, when I was invited to speak to the graduating seniors at the Science Convocation at McGill University in June 2003. I could assume that most of them were at least considering a career in scientific research. Like most scientists, I have had to give some thought to what works and what does not in research, and I couldn't resist the chance to burden an audience of future scientists with my opinions. I thought that the talk went pretty well, so I added a grandiose title and offered it to the venerable periodical *Nature,* which published it in November 2003.

When I received my undergraduate degree—about a hundred years ago—the physics literature seemed to me a vast, unexplored ocean, every part of which I had to chart before beginning any research of my own. How could I do anything without knowing everything that had already been done? Fortunately, in my first year of graduate school, I had the good luck to fall into the hands of senior physicists[1] who insisted, over my anxious objections,

1. Added note: They were David Frisch and Gunnar Källén, at the Bohr Institute in Copenhagen.

that I must start doing research, and pick up what I needed to know as I went along. It was sink or swim. To my surprise, I found that this works. I managed to get a quick Ph.D.—though when I got it I knew almost nothing about physics. But I did learn one big thing: that no one knows everything, and you don't have to.

Another lesson to be learned, to continue using my oceanographic metaphor, is that while you are swimming and not sinking you should aim for rough water. When I was teaching at the Massachusetts Institute of Technology in the late 1960s, a student told me that he wanted to go into General Relativity rather than the area I was working on, elementary particle physics, because the principles of the former were well known, while the latter seemed like a mess to him. It struck me that he had just given a perfectly good reason for doing the opposite. Particle physics was an area where creative work could still be done. It really was a mess in the 1960s, but since that time the work of many theoretical and experimental physicists has been able to sort it out, and put everything (well, almost everything) together in a beautiful theory known as the Standard Model. My advice is to go for the messes—that's where the action is.

My third piece of advice is probably the hardest to take. It is to forgive yourself for wasting time. Students in physics classes are only asked to solve problems that their professors (unless unusually cruel) know to be solvable. In addition, it doesn't matter if the problems are scientifically important—they have to be solved to pass the course. But in the real world, it's very hard to know which problems are important, and you never know whether at a given moment in history a problem is solvable.

At the beginning of the twentieth century, several leading physicists, including Lorentz and Abraham, were trying to work out a theory of the electron. This was partly in order to understand why all attempts to detect effects of Earth's motion through the ether had failed. We now know that they were working on the wrong problem. At that time, no one could have developed a successful theory of the electron, because quantum mechanics had not yet been discovered. It took the genius of Albert Einstein in 1905 to realize that the right problem on which to work was the effect of motion on measurements of space and time. This led him to the Special Theory of Relativity.

As you will never be sure which are the right problems to work on, most of the time that you spend in the laboratory or at your desk will be wasted. If you want to be creative, then you will have to get used to spending most of your time not being creative, to being becalmed on the ocean of scientific knowledge.

Finally, learn something about the history of science, or at a minimum the history of your own branch of science. The least important reason for this is that the history may actually be of some use to you in your own scientific work. For instance, now and then scientists are hampered by believing one of the oversimplified models of science that have been proposed by philosophers from Francis Bacon to Thomas Kuhn and Karl Popper. The best antidote to the philosophy of science is a knowledge of the history of science.

More importantly, the history of science can make your work seem more worthwhile to you. As a scientist, you're probably not going to get rich. Your friends and relatives probably won't understand what you're doing. And if you work in a field like elementary particle physics, you won't even have the satisfaction of doing something that is immediately useful. But you can get great satisfaction by recognizing that your work in science is a part of history.

Look back one hundred years, to 1903. How important is it now who was prime minister of Great Britain or Canada in 1903? What stands out as really important is that at McGill University, Ernest Rutherford and Frederick Soddy were working out the nature of radioactivity. This work of course had practical applications, but much more important were its cultural implications. The understanding of radioactivity allowed physicists to explain how the Sun and Earth's cores could still be hot after millions of years. In this way, it removed the last scientific objection to what many geologists and paleontologists thought was the great age of the Earth and the Sun. After this, Christians and Jews either had to give up belief in the literal truth of the Bible or resign themselves to intellectual irrelevance. This was just one step in a sequence of steps from Galileo through Newton and Darwin to the present that, time after time, has weakened the hold of religious dogmatism. Reading any newspaper nowadays is enough to show you that this work is not yet complete. But it is civilizing work, of which scientists are able to feel proud.

14

The Wrong Stuff

My work as a theoretical physicist is done with pen and paper, and so does not depend on large government grants for equipment. But theoretical research perishes without continual nourishment by new experimental data. In my own fields, elementary particle physics and cosmology, most of the data comes from expensive facilities— high energy accelerators many miles in circumference, huge tanks of fluid deep underground waiting to capture an occasional neutrino or dark matter particle or proton decay, telescopes many meters in diameter on mountaintops in Chile or Hawaii or West Texas, and telescopes and microwave receivers carried above the Earth's atmosphere by unmanned artificial satellites. It is always a struggle to get these facilities funded, and sometimes projects are cancelled after years of work and the expenditure of millions or billions of dollars. The traumatic and foolish cancellation in 1993 of the Superconducting Super Collider (described in my book *Dreams of a Final Theory*) has led to the loss of American leadership in high-energy physics research, and years of stunted progress in the field, which only now may be coming to an end with the completion of the Large Hadron Collider in Europe.

At the same time that physicists were struggling to keep the Super Collider funded, Congress was also considering appropriations for the International Space Station. The Space Station has cost over ten times as much as it would have taken to complete the Super Collider, and has contributed nothing of importance to scientific knowledge. Even so, when I testified in support of the Super Collider before a committee of the House of Representatives, I heard a congressman on the committee say that he saw how the

Space Station would contribute to understanding the universe, but he couldn't see that for the Super Collider.

Congress decided to keep the Space Station, and give up the Super Collider. This may be because NASA distributed Space Station contracts among almost every state. Perhaps if the Super Collider had been as expensive as the Space Station, and had spread its contracts throughout the United States, it would not have been cancelled. But I think that the Space Station had another thing working for it—a widespread fascination with seeing people in space. This silliness is often justified by talk of scientific research done by astronauts, but really has nothing to do with scientific work of any importance.

Then in January 2004 President Bush announced a new initiative in manned space flight that would send astronauts back to the Moon and then on to Mars. The cost was not announced, but it clearly would be enormous, much more even than the Space Station. Robert Silvers, editor of the *New York Review of Books,* knew I had strong views about manned space flight, because in an article on missile defense (number 7 in this collection) I had referred to the "idiocy" of the International Space Station. He asked me to write an article on the President's new initiative, and I leaped at the chance. This article was published in April 2004.

Ever since NASA was founded, the greater part of its resources have gone into putting men and women into space. On January 14 of this year, President Bush announced a "New Vision for Space Exploration" that would further intensify NASA's concentration on manned space flight. The International Space Station, which has been under construction since 1998, would be completed by 2010; it would be kept in service until around 2016, with American activities on the station from now on focused on studies of the long-term effects of space travel on astronauts. The manned spacecraft called the space shuttle would continue flying until 2010, and be used chiefly to service the space station. The shuttle would then be replaced by a new manned spacecraft, to be developed and tested by 2008. Between 2015 and 2020 the new spacecraft would be used to send astronauts back to the Moon, where they would live and work for increasing periods. We would then be ready for the next step—a human mission to Mars.

This would be expensive. The President gave no cost estimates, but John McCain, chairman of the Senate Commerce, Science, and Transportation Committee, has cited reports that the new initiative would cost between $170 billion and $600 billion. According to NASA briefing documents, the figure of $170 billion is intended to take NASA only up to 2020, and does not include the cost of the Mars mission itself. After the former President Bush announced a similar initiative in 1989, NASA estimated that the cost of sending astronauts to the Moon and Mars would be either $471 billion or $541 billion in 1991 dollars, depending on the method of calculation. This is roughly $900 billion in today's dollars. Whatever cost may be estimated by NASA for the new initiative, we can expect cost overruns like those that have often accompanied big NASA programs. (In 1984 NASA estimated that it would cost $8 billion to put the International Space Station in place, not counting the cost of using it. I have seen figures for its cost so far ranging from $25 billion to $60 billion, and the station is far from finished.) Let's not haggle over a hundred billion dollars more or less—I'll estimate that the President's new initiative will cost nearly a trillion dollars.

Compare this with the $820 *million* cost of recently sending the robots Spirit and Opportunity to Mars, roughly one thousandth the cost of the President's initiative. The inclusion of people inevitably makes any space mission vastly more expensive. People need air and water and food. They have to be protected against cosmic rays, from which we on the ground are shielded by the Earth's atmosphere. On a voyage to Mars, astronauts would be beyond the protective reach of the Earth's magnetic field, so they would also have to be shielded from the charged particles that are sent out by the Sun during solar flares. Unlike robots, astronauts will want to return to Earth. Above all, the tragic loss of astronauts cannot be shrugged off like the loss of robots, so any casualties in the use of the new spacecraft would cause costly delays and alterations in the program, as happened after the disastrous accidents to the *Challenger* shuttle in 1986 and to the *Columbia* shuttle in February 2003.

The President's new initiative thus makes it necessary once more to take up a question that has been with us since the first space ventures: What is the value of sending human beings into space? There is a serious conflict here. Astronomers and other scientists are generally skeptical of the value of manned space flight, and often resent the way it interferes with scientific research. NASA administrators, astronauts, aerospace contractors, and politicians typically find manned space flight just wonderful. NASA's Office of Space Science has explained that "the fundamental goal of the President's Vision is to advance US scientific, security, and economic interests through a robust space exploration program." So let's look at how manned space flight advances these interests.

Many Americans remember the fears for U.S. national security that were widely felt when the Soviets launched the unmanned *Sputnik* satellite in October 1957. These fears were raised to new heights in 1961, when the Soviet cosmonauts Yuri Gagarin and then Gherman Titov went into space. Titov's spacecraft made seventeen orbits around the Earth, three of them passing for the first time over the United States. The American reaction is described by Tom Wolfe in *The Right Stuff*:

> Once again, all over the country, politicians and the press seemed profoundly alarmed, and the awful vision was presented; suppose the cosmonaut were armed with hydrogen bombs and flung them as he came over, like Thor flinging thunderbolts. . . . Toledo disappears off the face of the earth . . . Kansas City . . . Lubbock. . . .

As it turned out, the ability to send rockets into space did have tremendous military importance. Ballistic missiles that travel above the Earth's atmosphere all but replaced bombers as the vehicle of choice for carrying Soviet or American nuclear weapons to an adversary's territory. Even in the nonnuclear wars of today, artificial satellites in orbit around the Earth play an essential part in surveillance, communications, and navigation. But these missiles and satellites are all unmanned. As far as I know there never has been a moment from Titov's flight to the present when the ability to put people into space gave any country the slightest military advantage.

I say this despite the fact that some military satellites have been put into orbit by the space shuttle. This could be done just as well and much more cheaply by unmanned rockets. It had been hoped that the shuttle, because reusable, would reduce the cost of putting satellites in orbit. Instead, while it costs about $3,000 a pound to use unmanned rockets to put satellites in orbit, the cost of doing this with the shuttle is about $10,000 a pound. The physicist Robert Park has pointed out that at this rate, even if lead could be turned into gold in orbit, it would not pay to send it up on the shuttle. Park could have added that in this case NASA would probably send lead bricks up on the shuttle anyway, and cite the gold in press releases as proof of the shuttle's value.

There doesn't seem to have been any reason for the use of the shuttle to take some military satellites into orbit other than that NASA has needed some way to justify the shuttle's existence. During the Carter administration, NASA explained to the deputy national security adviser that unless President Carter forced military satellite missions onto the space shuttle it would be the President who would be responsible for the end of the shuttle program, since the shuttle could never survive if it had to charge commercial users the real cost of space launches.

Similar remarks apply to the direct economic benefits of space travel. There is no doubt about the great economic value of artificial satellites in orbit around the Earth. Those that survey the Earth's surface give us information about weather, climate, and environmental change of all sorts, as well as warnings of military buildups and rocket launches. Satellites relay television programs and telephone conversations beyond the horizon. The Global Positioning System, which calculates the location of automobiles, ships, and planes, as well as missiles, relies on the timing of signals from satellites. But again, these are all unmanned satellites, and can be put into orbit most cheaply by unmanned rockets.

It is difficult to think of any direct economic benefit that can be gained by putting people into space. There has been a continuing effort to grow certain crystals in the nearly zero gravity on an orbiting satellite such as the International Space Station, or to make ultrapure semiconductor films in the nearly perfect vacuum in the wake of the space station. Originally President Reagan approved the space station in the expectation that eventually it could be run at a profit. Nothing of economic value has come of this, and these programs have now apparently been wisely abandoned in the President's new plans for the space station.

Lately there has been some talk of sending astronauts to mine the light isotope helium three on the Moon, where it has been deposited through billions of years of exposure of the Moon's surface to the solar wind. The point is that the more familiar thermonuclear reactions that use hydrogen isotopes as fuel produce large numbers of neutrons, which could damage surrounding materials and make them radioactive, while thermonuclear reactions involving helium three produce far fewer neutrons, and hence less radioactive waste. A thermonuclear reactor using helium three might also allow a more efficient conversion of nuclear energy to electricity, if it could be made to work.

Unfortunately, that is a big "if." One of the things that makes the development of thermonuclear power so difficult is the necessity of heating the fuel to a very high temperature so that atomic nuclei can collide with each other with enough velocity to overcome the repulsive forces between the electric charges carried by the nuclei. Helium nuclei have twice the electric charge of hydrogen nuclei, so the temperature needed to produce thermonuclear reactions involving helium three and hydrogen isotopes is much higher than the temperature needed for reactions involving hydrogen isotopes alone. So far, no one has been able to produce a useful, self-sustaining thermonuclear reaction using hydrogen isotopes. Until that is done, there seems little point in going to great expense on the Moon to mine a fuel whose use would make it even more difficult to generate thermonuclear power.

In his speech on January 14 President Bush emphasized that the space program produces "technological advances that have benefited all humanity." It is true that pursuing a demanding task like putting men on Mars can yield indirect benefits in the form of new technologies, but here too I think that unmanned missions are likely to be more productive. Trying to think of some future spin-off from space missions that would really benefit humanity, I find it hard to come up with anything more promising than the experience of designing robots that are needed for unmanned space missions.[1] This experience can help us in building robots that can spare humans from dangerous or tedious jobs here on Earth. Surprises are always possible, but I don't see how anything of comparable value could come out of developing the specialized techniques needed to keep people alive on space missions.

President Bush's presentation of his space initiative emphasized the scientific knowledge to be gained. Some readers of his speech may imagine astronauts on the shuttle or the space station peering through telescopes at planets or stars, or wandering about on the Moon or Mars making discoveries about the history of the solar system. It doesn't work that way.

There is no question that observatories in space have led to a tremendous increase in astronomical knowledge. To take just one example, in the early 1990s instruments on the Cosmic Background Explorer satellite made measurements of a faint background

---

1. Added note: When I made this remark in a talk at a computer consortium in Austin, someone in the audience pointed out that unmanned space missions may make an even greater contribution to technology by spurring the development of computer programs that can deal with unforeseen problems in real time.

of microwave radio static that had been discovered in 1965. This radiation is left over from a time when the universe in its present phase of expansion was only about four hundred thousand years old. The new data showed that the intensity of this radiation at various wavelengths is now just what would be emitted by opaque matter at a temperature of 2.725 degrees Celsius above absolute zero. It was the first time in the history of cosmology that anything had been measured to four significant figures. More important, the intensity of this radiation was found to be not perfectly uniform, but slightly lumpy. The observed intensity differs from one part of the sky to another by roughly one part in a hundred thousand for directions separated by a few degrees of arc. This amount of lumpiness is just what was expected, on the assumption that these variations in the cosmic microwave background arose from quantum fluctuations in the spatial distribution of energy in the very early universe, fluctuations that also eventually gave rise to the concentrations of matter—galaxies and clusters of galaxies—that astronomers see throughout the universe. There followed a decade of increasingly refined observations of the cosmic microwave background from mountaintops, balloons, and the South Pole, but the distorting effect of the Earth's atmosphere sets a limit to the precision that can be obtained with measurements from even the highest altitudes accessible to balloons. Finally these studies were dramatically advanced in 2002 by a remarkable new space mission, the Wilkinson Microwave Anisotropy Probe. After making repeated loops around the Moon to build up speed, this probe traveled out to a point in space known as L2, about a million miles from the Earth (four times farther than the Moon), in the direction opposite from the Sun. Anything placed at L2 orbits the Sun at just the speed needed to keep it at L2. There, in the cold quiet of interplanetary space, it was possible to map out the lumpiness of the cosmic microwave radiation background to an unprecedented level of accuracy. The comparison of these measurements with theory has confirmed our general ideas about the emergence of fluctuations in the very early universe; it has shown that the universe now consists of about 4 percent ordinary atoms, about 23 percent dark matter of some exotic type that does not interact with radiation, and the rest some sort of mysterious "dark energy" having negative pressure; and it has given the age of the universe in its present phase as between 13.5 billion and 13.9 billion years.

Exciting research, of which NASA may justly feel proud. Research of this sort has made this a golden age for cosmology. But neither the Cosmic Microwave Background Explorer nor the Wilkinson Microwave Anisotropy Probe had any astronauts aboard. People were not needed. On the contrary, through their movements and body heat they would have fouled up these measurements, as well as greatly increasing the cost of these missions. The same is true of every one of the space observatories that have expanded our knowledge of the universe through observations of ultraviolet light, infrared light, X-rays, or gamma rays from above the Earth's atmosphere. Some of these observatories were taken into orbit by the shuttle, while others (including the Cosmic Microwave Background Explorer and the Wilkinson Microwave Anisotropy Probe) were carried up by unmanned rockets, as all of them could have been.

The Hubble Space Telescope is a special case. Like the other orbiting observatories, the Hubble operates under remote control, with no people traveling with it. But unlike these other observatories, the Hubble was not only launched by the shuttle but has also been serviced several times by astronauts brought up to its orbit by the shuttle. The Hubble has made a great contribution to astronomy, one that goes way beyond taking gorgeous color photos of planets and nebulae. Most dramatically, teaming up with observatories on the ground, the Hubble carried out a program of measuring the distances and velocities of faraway galaxies. In 1998 these measurements revealed that the expansion of the universe is not being slowed down by the mutual gravitational attraction of its matter, as had been thought, but is rather speeding up, presumably in response to the gravitational repulsion of the dark energy I mentioned earlier. [Dark energy is discussed in more detail in chapter 5 in this collection.] The Hubble may have given NASA its best argument for the scientific value of manned space flight.

But like the other space observatories, the Hubble Space Telescope could have been carried into orbit by unmanned rockets. This would have spared astronauts the danger of shuttle flights, and it would have been much cheaper. Riccardo Giacconi, the former director of the Space Telescope Science Institute, has estimated that by using unmanned rockets instead of the space shuttle, we could have sent up seven Hubbles without increasing the total mission cost. It would then not have been necessary to service the Hubble; when design flaws were discovered or parts wore out, we could just have sent up another Hubble.

What about the scientific experiments done by astronauts on the space shuttle or the space station? Recently I asked to see the list of experiments that NASA assigned to the astronauts aboard the *Columbia* space shuttle on its last flight, which ended tragically when the shuttle exploded during reentry. It is sad to report that it is not an impressive list of experiments. Roughly half had to do with the effect of the space environment on the astronauts. This at least is a kind of science that cannot be done without the presence of astronauts, but it has no point unless one plans to put people into space for long periods for some other reason.

Of the other half of the *Columbia*'s experiments, a large fraction dealt with the growth of crystals and the flow of fluids in nearly zero gravity, old standbys of NASA that have neither illuminated any fundamental issues of science nor led to any practical applications. It is always dangerous for a scientist in one field to try to judge the value of work done by specialists in other fields, but I think I would have heard about it if anything really exciting was coming out of any of these experiments, and I haven't. Much of the "scientific" program assigned to astronauts on the space shuttle and the space station has the flavor of projects done for a high school science talent contest. Some of the work looks interesting, but it is hard to see why it has to be done by people. For instance, there was just one experiment on *Columbia* devoted to astronomy, a useful measurement of variations in the energy being emitted from the Sun. The principal investigator tells me that the only intervention of the astronauts consisted of turning the apparatus on and then turning it off.

Looking into the future, we need to ask, what scientific work can be done by astronauts on Mars? They can walk around and look at the terrain, and carry out tests on rocks, looking for signs of water or life, but all that can be done by robots. They can bring back rock samples, as the *Apollo* astronauts did from the Moon, but that too can be done by robots. Samples of rocks from the Moon were also brought back to Earth by unmanned Soviet lunar missions. It is sometimes said that the great disadvantage of using robots in a mission to Mars is that they can only be controlled by people on Earth with a long wait (at least four minutes) for radio signals to travel each way between the Earth and Mars. That would indeed be a severe problem if the robots were being sent to Mars to negotiate with Martians, but not much is happening there now, and I don't see why robots can't be left to operate with only occasional intervention from Earth. Any marginal advantage that astronauts may have over robots in exploring Mars would be more than cancelled by the great cost of manned missions. For the cost of putting a few people in a single location on Mars, we could have robots studying many different landscapes all over the planet.

Many scientists and some NASA administrators understand all this very well. I have frequently been told that it is necessary publicly to defend programs of manned space flight anyway, because the voters and their elected representatives only care about the drama of people in space. (Richard Garwin has reminded me of the old astronauts' proverb "No bucks without Buck Rogers.") It is hoped that while vast sums are being spent on manned space flight missions, a little money will be diverted to real science. I think that this attitude is self-defeating. Whenever NASA runs into trouble, it is science that is likely to be sacrificed first. After NASA had pushed the *Apollo* program to the point where people stopped watching lunar landings on television, it cancelled *Apollo 18* and *19*, the missions that were to be specifically devoted to scientific research.

It is true that the administration now projects a 5 percent increase per year in NASA's funding for the next three years. So far, funding is being maintained for the next large space telescope, and is being increased for some other scientific programs, including robotic missions to the planets and their moons. But we can already see damage to programs that are not related to exploration of the solar system, and especially to research in cosmology. Studying the origin of the planets is interesting, but certainly not more so than studying the origin of the universe.

Two days after President Bush presented his new space initiative, NASA announced that the planned shuttle mission to service Hubble in 2006 would be cancelled. This mission would have replaced gyroscopes and batteries that are needed to extend Hubble's life into the next decade, and it would have installed two new instruments (which have already been built, at a cost of $167 million) to extend Hubble's capacities. One of these instruments would have allowed Hubble to survey the sky in infrared and ultraviolet light, revealing much about the formation of the earliest stars and galaxies. The other was an ultraviolet spectrograph, which would have explored intergalactic matter in the early universe. Using older instruments, Hubble would also have pushed the program of measuring distances and velocities of galaxies to greater distances, mapping out the dark energy that is accelerating the expansion of the universe. Instead, in about three years, when the Hubble gyroscopes can no longer point the telescope accurately, it will cease operation. This will leave us with no large space telescope until 2011 at the earliest. Eventually, before the slight drag of the Earth's atmosphere at its altitude can bring the Hubble down, an unmanned rocket will be sent up to the Hubble to take it out of orbit and deposit it harmlessly into the ocean. Part of the increase in NASA's spending for science will be about $300 million for destroying Hubble.

NASA's stated reason for terminating the Hubble while continuing work on the space station is that it is more dangerous for the shuttle to go up to Hubble than to the space station. Supposedly, if the astronauts on the shuttle find that damage has been done to the shuttle's protective tiles during launch, they could wait in the space station for a rescue, while this would not be possible during a mission to the Hubble. But there are many other dangers to astronauts that are the same whether the shuttle is going to the space station or the Hubble Space Telescope. Among these is an explosion during launch, like the one that destroyed the *Challenger* shuttle in 1986. The *New York Times* Web site has carried a report from an anonymous NASA engineer who challenges NASA's statement that a shuttle flight to Hubble would be more risky than a flight to the space station. He or she points out that the shuttle would be less exposed to micrometeoroids and orbital debris at the altitude of Hubble than at the lower altitude of the space station. Even if one considers only the possibility of damage during launch to the shuttle's protective tiles, there may not be much difference in the risks of going to Hubble or the space station. The *Columbia* Accident Investigation Board discussed this safety problem, but it recommended that NASA develop the ability to repair the shuttle's tiles in space, whether or not it is docked to the space station, and it did not conclude that the Hubble had to be abandoned. To be reasonably sure of rescuing the astronauts even if it turns out that damage to the shuttle can't be repaired in space, it could be arranged at some extra cost that when one shuttle goes up to service Hubble, the other shuttle will be kept ready on the ground. For instance, the Hubble servicing mission could be scheduled just before one of the planned missions to the space station.

In response to pressure from Congress and the scientific community, NASA has agreed to reconsider this decision. I don't know enough about questions of safety to judge this issue myself, but I share the widespread suspicion that Hubble is being sacrificed to save funds for the President's initiative, and in particular in order to reserve all flights on the shuttle's limited schedule for the one purpose of taking astronauts to and from the space station.

Perhaps because of its timing, the Hubble decision attracted great public attention, but there are other recent NASA decisions that have nothing to do with safety, and that therefore give clearer evidence of the willingness of NASA and the administration to sacrifice science to save money for manned space flight. In January 2003, after several years of scientists' making difficult decisions about their priorities, NASA announced a new initiative, called Beyond Einstein, to explore some of the more exotic phenomena predicted by Einstein's General Theory of Relativity. This includes a satellite (to be developed jointly with the Department of Energy) that would look at many more galaxies at great distances, in order to uncover the nature of the dark energy by finding whether its density has been changing as the universe expands. Equally important for cosmology, there would be another probe that would study the polarization of the cosmic microwave background to find indirect effects of gravitational waves from the early universe. (Gravitational waves bear the same relation to ordinary gravity that light waves bear to electric and magnetic fields; they are self-sustaining oscillations in the gravitational field, which propagate through empty space at the speed of light.)

Beyond Einstein also includes another satellite dedicated to searching for black holes, and two larger facilities. One is an array of X-ray telescopes called Constellation-X, which would observe matter falling into black holes. The other is called LISA, the Laser Interferometer Space Antenna. This "antenna" would consist of three unmanned spacecraft in orbit around the Sun, separated from each other by about three million miles. Changes in the distances between the three spacecraft would be continually measured with a precision better than a millionth of an inch by combining laser beams passing between them. These exquisite measurements would be able to reveal the presence of gravitational waves passing through the solar system. LISA would have enough sensitivity to detect gravitational waves produced by stars being torn apart as they fall into black holes or by black holes merging with each other, events we can't see with ordinary telescopes. NASA has another particularly cost-effective older program called Explorer, which has supported small and mid-sized observatories like the Cosmic Background Explorer and Wilkinson Microwave Anisotropy Probe.

Alas, NASA's Office of Space Science has now announced that the Beyond Einstein and Explorer programs "do not clearly support the goals of the President's Vision for space exploration," so their funding has been severely reduced. Funding for the three smaller Einstein missions has been put off for five years; LISA will be deferred for a year or more; Explorer will be reduced in scope for the next five years; and no proposals for new Explorer missions will be considered for one or two years. None of this damage is irreparable, but spending on the President's "New Vision" has barely begun. These deferrals, along with the end of Hubble servicing, are warnings that as the Moon and Mars missions absorb more and more money, the golden age of cosmology is going to be terminated, in order to provide us with the spectacle of people going into space for no particular reason.

When advocates of manned space flight run out of arguments for its contribution to "scientific, security, and economic interests," they invoke the spirit of exploration, and talk of the Oregon Trail (Bush I) or Lewis and Clark (Bush II). Like many others, I am not immune to the excitement of seeing astronauts walking on Mars or the Moon. We have walked on Mars so often in our reading—with Dante and Beatrice, visiting the planet of martyrs and heroes; with Ray Bradbury's earthmen, finding ruins and revenants of a vanished Martian civilization; and more recently with Kim Stanley Robinson's pioneers, transforming Mars into a new home for humans. I hope that someday men and women will walk on the surface of Mars. But before then, there are two conditions that will need to be satisfied.

One condition is that there will have to be something for people to do on Mars that cannot be done by robots. If a few astronauts travel to Mars, plant a flag, look at some rocks, hit a few golf balls, and then come back, it will be a thrilling moment, but then, when nothing much comes of it, we will be left with a sour sense of disillusionment, much as happened after the end of the *Apollo* missions to the Moon. Perhaps after sending more robots to various sites on Mars something will be encountered that calls for direct study by humans. Until then, there is no point in people going there.

The other necessary condition is a reorientation of American thinking about government spending. There seems to be a general impression that government spending harms the economy by taking funds from the private sector, and therefore must always be kept to a minimum. Unlike what is usually called "big science"— orbiting telescopes, particle accelerators, genome projects—sending humans to the Moon and Mars is so expensive that, as long as the public thinks of government spending as parasitic on the private economy, this program would interfere with adequate support for health care, homeland security, education, and other public goods, as it has already begun to interfere with spending on science.

My training is in physics, so I hesitate to make pronouncements about economics; but it seems obvious to me that for the government to spend a dollar on public goods affects total economic activity and employment in just about the same way as for government to cut taxes by a dollar that will then be spent on private goods. The chief difference is in the kind of goods produced by the economy—public or private. The question of what kind of goods we most need is not one of economic science but of value judgments, which anyone is competent to make. In my view the worst problem facing our society is not that there is a scarcity of private goods—food or clothing or SUVs or consumer electronics—but rather that there are sick people who cannot get health care, drug addicts who cannot get into rehabilitation programs, ports vulnerable to terrorist attack, insufficient resources to deal with Afghanistan and Iraq, and American children who are being left behind. As Justice Holmes said, "Taxes are what we pay for civilized society." But as long as the public is so averse to being taxed, there will be even less money either to ameliorate these societal problems or to do real scientific research if we spend hundreds of billions of dollars on sending people into space.

In the foregoing, I have taken the President's space initiative seriously. That may be a mistake. Before the "New Vision" was announced, the administration was faced with the risk of political damage from a possible new fatal shuttle accident like the *Columbia* disaster less than a year earlier. That problem could be eased by cancelling all shuttle flights before the 2004 presidential election, and allowing only enough flights after that to keep building the space station. The space station posed another problem: no one was excited any more by what had become the Great Orbital Turkey. While commitments to domestic contractors and international partners protected it from being immediately scrapped, its runaway costs needed to be cut. But just cutting back on the shuttle and the space station would be too negative, not at all in keeping with what might be expected from a President of Vision. So, back to the Moon, and on to Mars! Most of the huge bills for these manned missions would come due after the President leaves office in 2005 or 2009, and the extra costs before then could be covered in part by cutting other things that no one in the White House is interested in anyway, like research on black holes and cosmology. After the end of the President's time in office, who cares? If future presidents are not willing to fund this initiative, then it is they who will have to bear the stigma of limited vision. So, looking on the bright side, instead of spending nearly a trillion dollars on manned missions to the Moon and Mars we may wind up having spent only a fraction of that on nothing at all.

Since this article was published in April 2004, my fears—that the cost of the President's initiative would stifle programs of astronomical research by unmanned satellites—have largely come true. Of the various space-borne research facilities that were supposed to be supported by the Beyond Einstein program, none have been launched, and only one, a mission to study dark energy, seems to be going ahead. In the Explorer program, the Swift observatory was essentially complete at the time President Bush announced his Moon-Mars mission in January 2004, and it was launched soon after. Its observations of gamma rays, X-rays, and ultraviolet radiation have been highly successful. But since 2004, no new astronomical Explorer missions have been launched.

There are some signs that public opinion may be changing. Under intense political pressure, a new NASA director in the Bush administration, Michael Griffin, agreed to send a space shuttle to service the Hubble telescope; this was accomplished without mishap in May 2009. Griffin is not unfriendly to science, but he is an enthusiastic supporter of manned space flight. (I had lunch with Griffin about a year ago, and handed him a copy of the issue of the *New York Review of Books* containing the foregoing article. He was kind enough to say that he would enjoy reading it on his flight back to Washington. I warned him that he wouldn't.) A new President has just taken office, and Griffin has resigned as director of NASA. We can hope that the Obama administration will cancel the folly of the Moon-Mars mission, but it is too soon to tell.

15

A Turning Point?

As a contribution to the nation's debate during the 2004 presidential election, the *New York Review of Books* invited fourteen previous contributors to submit brief articles on election issues. From articles I had published, Robert Silvers of the *New York Review* knew that I had strong views regarding the Bush administration initiatives in missile defense, nuclear weapons development, and manned space flight, and asked me to be one of the contributors.

In the following essay I took the occasion also to express my views on several other issues, on which I had not previously written. The issue of the *New York Review* containing the symposium "The Election and America's Future" was published just before the November 2004 election. Somehow President Bush was reelected anyway.

Anyone speeding down a freeway in the wrong direction will naturally begin to think of turning points. I don't know if the 2004 election will be a turning point for America, but here are some thoughts about the direction toward which we ought to turn, with emphasis on policies that are abhorrent to Republicans and also rarely espoused by Democrats.

Here in Texas, I meet a good many economic conservatives. Most are nice people, not naturally inclined toward grinding the faces of the poor. But when I argue that we need to elect a president and Congress willing to raise taxes and spend more on searching for nuclear weapons in container ships, decreasing public school class sizes, subsidizing secular education in Islamic coun-

tries, rebuilding Afghanistan, expanding scientific research, help-
ing the Russians to control their stocks of fissionable material, and
providing universal access to health care (perhaps by offering
Medicare to anyone who wants it), my conservative friends claim
that we can afford these things only if we first "grow the economy" by cutting taxes. They (and many liberals) are in the grip of
an economic fallacy, that a dollar spent on consumer goods stimu-
lates the economy more than a dollar spent on public goods. They
actually believe that holding government spending down by not
hiring enough policemen and patent inspectors is the way to fight
unemployment!

Ironically, cutting taxes on high incomes and large legacies has
weakened a unique aspect of American life that conservatives
ought to prize. By making the deduction for charitable donations
less attractive, tax cuts have hurt private foundations, universities,
museums, orchestras, hospitals, and churches, which although in-
directly supported by government through tax deductions for
donors, nevertheless operate without government control.

We must restore steeply progressive income taxes and cut pay-
roll taxes, in order to reverse the growth of poverty and narrow
the gap between the real net hourly earnings of most people and
the richest few. Even more important is keeping the inheritance tax
and closing its loopholes, to inhibit development of an aristocracy
of inherited wealth. To restore the dignity as well as the prosperity
of working people, Congress needs to give workers some legal
right to their jobs, prohibit the replacement of striking workers,
and repeal the worst features of the Taft-Hartley law. No one
should imagine that the economy necessarily will be healthy if only
unemployment rates can be lowered. No class of Americans has
ever had unemployment as low as blacks in the antebellum South.

One other kind of tax needs to be increased, the tax on gasoline.
We need to get gasoline prices up permanently without sending
more money to the Saudis, to provide an incentive for fuel efficiency and alternatives. The worst possible policy would be to try
to hold gasoline prices down by drilling in Alaska. A few decades

from now, when the only large conventional oil reserves left in the world will be in the Middle East, the oil in Alaska can be a precious asset, our ultimate strategic reserve.

Part of the cost of the public goods we need can be borne by cutting back on those we don't need. The President's vastly expensive "New Vision" for manned space flight serves no economic, military, or scientific purpose. The antimissile defense now being deployed at great cost will have only a dubious effectiveness against the most unlikely nuclear threats, and concededly no effectiveness at all against the one peril that can destroy our country, a massive Russian missile launch by mistake.

John Kerry's public statements have not revealed how he would change direction on these issues. Maybe that is good strategy—what do I know about how to win an election? Of one thing I am sure: we can look for no help from George Bush.

President Bush's reelection would be disastrous in another respect. The present Supreme Court has attacked the constitutional powers of Congress, striking down legislation that would protect individuals against unconstitutional state action. The vacancies on the Court that are likely to open soon create an opportunity to reverse these decisions. Four more years of a Bush administration will tip the balance of the Court toward extremist justices like Antonin Scalia and Clarence Thomas, whom Bush especially admires.

After all this, you would think that I would have no doubt about my vote in November, but I have one remaining concern that might keep me from voting for Kerry. Somehow there has grown up a correlation between liberalism and anti-Zionism in both Europe and America: a tendency for the same politicians, academics, performers, and journalists who take a liberal stand on domestic issues reflexively to take the Arab side in disputes between Arabs and Israelis. Kerry's statements and voting record show no signs of anti-Zionism, with just one exception known to me, his speech at the Council on Foreign Relations naming James Baker of all people as someone he might send to make peace between Israel and the Palestinians—a possibility he subsequently rejected.

Nevertheless, I can't help worrying about the foreign policy of a liberal administration if Kerry is elected. This concern is deepened by the fear that, as radical Islamic terrorism continues to plague us, there will be a growing temptation to appease Muslims either by withdrawing support for Israel, or by making complete withdrawal from the West Bank a condition for this support, leaving Israel vulnerable to the sort of attack launched by Arab states in 1948, 1967, and 1973. Yielding to this temptation would weaken the cause of secular democracy, and permanently stain our country's honor. But, hoping for the best, I expect to vote for Kerry anyway.

16

About Oppenheimer

This is a review of *Oppenheimer: Portrait of an Enigma* (I. R. Dee, Chicago, 2004), by the physicist and author Jeremy Bernstein. I agreed to review the book for *Physics Today* partly because I always enjoy Bernstein's writing, and also because at one point a story about his subject, J. Robert Oppenheimer, played an important part in how I thought about my own career.

Deciding to be a scientist is a little like taking the veil. One may feel that scientific discovery has a transcendent importance, but at the same time one is generally resigning oneself to working outside the great world of public affairs. This is especially true for a theorist, who usually works alone or with one or two collaborators, and doubly so for someone like myself who works in elementary particle theory and cosmology, fields that are unlikely to yield discoveries of immediate practical importance. Since high school I have never had any doubt that I wanted to be a theoretical physicist, but at the same time I felt a certain wistfulness about some of the aspects of life that I would be missing.

Then sometime in the late 1940s, when I was still in high school, I saw an issue of some magazine, probably *Life,* with Oppenheimer on the cover. I read that he was engaged in profound research on elementary particles and collapsing stars, was a student of Sanskrit poetry, and through his leadership of the Manhattan Project at Los Alamos had done as much as any other American to end World War II. Wow! I saw that one didn't have to renounce the world when one took on the vocation of theoretical physics.

In fact, though I have never for a moment regretted my decision to become a physicist, my life has not been so worldly, certainly nothing like Oppenheimer's. Still, there was a moment in high school when learning about Oppenheimer greatly bucked me up.

The following review of Bernstein's book was published in the January 2005 issue of *Physics Today.*

Jeremy Bernstein is as good a writer as you can find among scientists. I would look forward to reading any biography written by Bernstein, even if the subject were Millard Fillmore, or Liberace. All the better if the subject is J. Robert Oppenheimer. He was one of the very few American theorists who participated in the quantum mechanics revolution in the 1920s; and in the 1930s he pioneered the theory of neutron stars and black holes. Though not in a class with theorists like Paul Dirac or Wolfgang Pauli, in his day in the 1930s Oppenheimer was one of the leaders of theoretical physics.

But of course it is as a public figure that Oppenheimer is so interesting. Under his direction, the secret laboratory at Los Alamos designed and built the nuclear weapons that ended World War II. After the war, he became the head of the Institute for Advanced Studies, in the glory days when the Institute housed Kurt Gödel, George Kennan, John von Neumann, and Hermann Weyl, as well as Albert Einstein. As chairman of the General Advisory Committee of the Atomic Energy Commission, Oppenheimer continued to advise on nuclear policy at the highest levels of government. Then, after a hearing rivaling the trials of Saint Joan and Charles I for drama, his security clearance was revoked.

All this made Oppenheimer a world celebrity. I remember that in 1962 my wife and I were sitting in a café in Geneva, during a break in the "Rochester" High Energy Physics Conference then in town. Looking at the other café patrons, we decided that they must be diplomats—they spoke languages we couldn't identify, and they were much too well dressed to be physicists. For a moment I felt that, although I loved physics, in choosing a career in research I had given up the glamour of the great world of national and international affairs. Then Oppenheimer came in. He stopped at our table, and chatted with me for a few minutes about some of the talks at the conference. After he walked away, one of the diplomats, wearing a gorgeous tarboosh and fez, came over and said "Pahdon me, sah, but was that Doctah Oppenheimah?" My fit of self-pity passed; I didn't have a diplomatic passport, but at least I knew Oppenheimer.

Bernstein's new book is a splendid brief biography of Oppenheimer, originally intended as a *New Yorker* profile. Bernstein is very good at describing Oppenheimer's strengths. Oppenheimer had the ability to understand anything—all the work being done in every department at Los Alamos, and then after the war everything going on at the frontier of theoretical physics. When I was a graduate student at Princeton I used to go over to Building E of the Institute to attend physics seminars. Oppenheimer always sat in the front row, asking questions that demonstrated that he knew as much about the speaker's subject as the speaker. Of course he was showing off, but no one else could have gotten away with it. He *did* know as much as the speaker.

Oppenheimer was wide open to new ideas, and did what he could to advance the work of younger theorists whose ideas impressed him. Oppenheimer recruited a stellar group of new members to the Institute, including the theoretical physicists Freeman Dyson, Bram Pais, Frank Yang, and (briefly) T. D. Lee. Until reading Bernstein's book I hadn't known that as early as 1943, Oppenheimer had tried to get Berkeley to recruit Richard Feynman, a move opposed for some reason by the department chairman, Raymond Birge. As Bernstein says, relations between Birge and Oppenheimer were strained. Bernstein may not know of their later reconciliation. In 1966 Oppenheimer, then dying of cancer, made a last trip to Berkeley. After a dinner at the Faculty Club, Birge rose to salute Oppenheimer. You have to imagine Birge as I saw him then—a little old man in a dark three-piece suit, with a bald, rather square head and a formal manner—the last person in the world from whom you would expect much in the way of emotional display. Birge described what Oppenheimer had done for Berkeley, for the nation, and for physics, paused briefly, and ended with the words "And I love him."

Bernstein is so good at making a page-turner out of a scientific advance that I wish he had told the story of Oppenheimer's role in grappling with the problem that calculations of energies and probabilities in quantum field theory frequently give results that are infinite. Oppenheimer had actually been the first to encounter one of these awkward infinities in a calculation of the effect of emitting and reabsorbing photons on the energy levels in atoms—essentially the same phenomenon that later became known as the Lamb shift. In a reaction to these infinities and to certain anomalies in cosmic ray showers, Oppenheimer became a leading advocate of the idea that there is something fundamentally wrong with quantum electrodynamics (the quantum field theory of photons, electrons, and antielectrons), that it simply can't be trusted to describe physical processes at distance scales much less than 10^{-13} centimeters. (His views are quoted by Robert Serber in *The Birth of Particle Physics,* ed. L. Brown and L. Hoddeson, Cambridge University Press, 1983.) Nevertheless, when it was discovered after the war that quantum electrodynamics gave perfectly sensible finite results if properly interpreted, Oppenheimer accepted this new view, and hired Dyson, who gave the clearest exposition of how to deal with these infinities, as a permanent member of the Institute.

Oppenheimer never used his position of leadership in physics research to play the part of a mandarin, who tries to control what other physicists do. In this, he presents a sharp contrast to Werner Heisenberg, a greater physicist, but one who did what he could after the war to force German physicists to work on his own ideas. I am convinced that one of the reasons that the United States was successful in developing nuclear weapons during the war while Germany was not is that we had Oppenheimer, while they had Heisenberg.

Bernstein is even better in describing Oppenheimer's weaknesses. He offers an interesting explanation why Oppenheimer, who had never organized anything more complicated than a camping trip, was tapped by General Leslie Groves in 1942 to lead the secret nuclear weapon laboratory at Los Alamos, rather than someone like E. O. Lawrence (the inventor of the cyclotron), who was accustomed to running a scientific laboratory. It is that Groves sensed in Oppenheimer a certain pliability, missing in Lawrence, that would allow Groves to have his way in running things. This pliability surfaced in Oppenheimer's willingness to have himself and all the other scientists at Los Alamos inducted into the army. Fortunately the idea was dropped. As Bernstein notes, it would have been a disaster to try to put characters like Richard Feynman in uniform. Oppenheimer did other peculiar things. He made a gratuitous and false comment to a security officer that Bernard Peters was a communist, and in his own security hearing he carelessly chose an attorney who had little experience in litigation.

About that security hearing, I agree with Bernstein that Oppenheimer was not only loyal (which was acknowledged in the hearing) but also was not a security risk. There has been an argument recently in the pages of the *New York Review of Books,* between Gregg Herken, who claims that Oppenheimer was a communist in the 1930s and early 1940s, and Daniel Kevles, who denies it. It is a fuzzy question. There is no question that Oppenheimer had friends who were communists, and for a while before he went to Los Alamos, Oppenheimer belonged to a small group of Berkeley faculty that, depending on how you look at it, can be regarded as a Communist Party cell or a left-wing coffee klatsch. As far as I know, Oppenheimer never took orders from any communist organization. There is certainly no evidence that he ever leaked anything to anyone. On the contrary, when his friend Haakon Chevalier tried to get Oppenheimer to pass on information about the Manhattan Project to George Eltenton, who had contacts at the Soviet consulate, Oppenheimer voluntarily reported Eltenton's name to a security officer, and although he tried at first to shield Chevalier, he eventually complied with an order from Groves and revealed Chevalier's role. (He said that Chevalier had approached him through Oppenheimer's brother Frank, a pointless lie that hurt him at his security hearing.) An FBI wiretap of the communist agent Steve Nelson heard Nelson complaining that Oppenheimer would not divulge any information. Furthermore, whatever sympathy Oppenheimer may have felt earlier for the Soviet Union, he lost it during the war, and certainly had none in the 1950s. In her journal, Chevalier's wife recalls Oppenheimer telling Chevalier during the war that the Soviet Union was not to be trusted. The information about Oppenheimer's communist contacts that was dragged into the 1954 hearings had already been known when Oppenheimer had been cleared in 1947. Oppenheimer lost his clearance not because he was a security risk, but because Edward Teller and the Air Force did not like the advice he was giving about nuclear-powered airplanes and the hydrogen bomb.

As Bernstein describes it, Oppenheimer's early life may have left him permanently unsure of himself. After a rather sheltered well-off childhood, Oppenheimer went to Harvard at a time when the university was trying to cut down on the number of Jewish students, and found himself pretty much a recluse there. Bernstein quotes a letter home in which Oppenheimer gushes about his writing and reading and ends by wishing he were dead. As Bernstein rightly says, "The whole tone makes one's flesh creep." Oppenheimer's postgraduate time at the University of Cambridge was worse, and he suffered a nervous breakdown. His life became happier only when he met Max Born, and found his vocation as a theoretical physicist.

Even his beginnings as a physicist left Oppenheimer with some unhappy memories. I remember that once as a postdoc at Columbia I was invited to give a talk at the Institute on some of my recent work. My talk was terribly formal, bristling with mathematical complications. Oppenheimer interrupted me, and said that I reminded him of himself at my age. I stupidly blurted out "Thank you," to which he gravely replied, "It wasn't a compliment."

17

Einstein's Search for Unification

1905 was the *annus mirabilis* of Albert Einstein. In that one year, Einstein introduced the idea that light comes in discrete bundles of energy, and used it to understand the photoelectric effect; he explained that the zigzag "Brownian" motion of small particles suspended in fluids is a consequence of collisions with fluid molecules, providing evidence for the atomic nature of matter; he invented the Special Theory of Relativity; and he inferred from it the relation between mass and energy. The centenary of these achievements was celebrated by declaring 2005 the World Year of Physics. As part of the celebration, a book, *Einstein—A Hundred Years of Relativity* (Palazzo Editions, Bath, England, 2005), was prepared by Andrew Robinson, in association with the Albert Einstein Archives at the Hebrew University in Jerusalem. In addition to Robinson's comprehensive text, the book includes a preface by Freeman Dyson, articles on special topics by Arthur Clarke, Philip Glass, and Stephen Hawking, among others, and a wealth of wonderful photographs. I was asked to contribute something on Einstein's search for a unified field theory. The following article is my contribution to Robinson's book, with the addition of a footnote that I hope will clarify some of the ideas discussed.

Some time in the 1970s I received a letter from a San Francisco dealer in rare books and manuscripts. Enclosed was a photocopy of a scientific article by Einstein, hand-written in German. The dealer asked me to read the article, and give him advice about its importance. I am no Einstein scholar, and I can read German only with great difficulty, but a quick look at the article showed that it was another one of Einstein's many attempts in the 1930s and 1940s to find a unified theory of the electromagnetic and gravitational fields. I wrote to the book dealer and told him that, while anything in Einstein's own hand must be of value, this particular paper was of no great importance in the history of science.

Albert Einstein was one of the greatest scientists of all time, a peer of Galileo, Newton, and Darwin, and certainly the leading physicist of the twentieth century. It is not surprising that, after his scientific triumphs in the years from 1905 to 1925, he would turn his attention to the search for a unified theory of electromagnetism and gravitation. The greatest advances in the history of physics have been marked by the discovery of theories that gave a unified explanation of phenomena that had previously seemed unrelated. In the seventeenth century, Isaac Newton had unified celestial and terrestrial physics, showing that the same force of gravitation that causes apples to fall to the ground also holds the Moon in its orbit around the Earth and the Earth and planets in their orbits around the Sun. In the nineteenth century, James Clerk Maxwell had unified the phenomena of electricity and magnetism, by realizing that just as oscillating magnetic fields produce electric forces, so also oscillating electric fields produce magnetic forces, and had discovered that light was nothing but a self-sustaining oscillation of both electric and magnetic fields. In 1915 Einstein himself, in his General Theory of Relativity, had shown that gravitation was but an aspect of the geometry of space and time. After that brilliant success, the obvious next problem was to find a theory that gave a unified account of gravitation and electromagnetism. Tragically, Einstein spent almost all of the last thirty years of his life pursuing this aim, not only without success, but without leaving any significant impact on the work of other physicists.

In this work Einstein followed two main lines of attack.

First, Einstein was attracted by an idea in a 1921 paper by the mathematician Theodor Kaluza, that electromagnetism could be understood as an aspect of gravitation in five rather than the usual four dimensions of space and time. It was clear from Einstein's development of General Relativity that gravitation in any number of dimensions should be described by a symmetric square array g_{MN}. (M labels the rows and N labels the columns of the array; "symmetric" means that $g_{MN} = g_{NM}$.) The entries g_{MN} in this array are fields—that is, they are numbers whose values depend on position in space and time. This array of fields also describes the geometry of space-time, and hence is known as the *metric*. In the usual four space-time dimensions, M and N run over the labels 1, 2, and 3 that are conventionally used to distinguish the three directions of ordinary space, plus a label 0 used to distinguish the dimension of time. That is, gravitation is described by 10 fields: g_{11} , g_{22} , g_{33} , g_{00}, $g_{12} = g_{21}$, $g_{23} = g_{32}$, $g_{10} = g_{01}$, $g_{20} = g_{02}$, and $g_{30} = g_{03}$.[1] In five dimensions M and N run over the labels 1, 2, 3, and 0, plus a label 5 for the fifth dimension. Kaluza proposed that the part of g_{MN} in which both M and N run only over 1, 2, 3 and 0 represents the gravitational field observed in four space-time dimensions, just as in Einstein's theory; that the four new quantities g_{51}, g_{52}, g_{53}, and g_{50} (which because the metric is symmetric are equal, respectively, to g_{15}, g_{25}, g_{35}, and g_{05}) form the "vector potential" that is often used as an alternative description of the electromagnetic field in conventional theories of electrodynamics; and that g_{55} is a new field representing some sort of matter. Indeed, with this interpretation of the fields g_{MN}, if one arbitrarily assumes that these fields are independent of position in the fifth dimension, then it turns out that the obvious extension of the field equations of General Relativity to five dimensions yields the equations of General Relativity for the gravitational field and the Maxwell equations for the electromagnetic field, both in four dimensions.

1. Added note: This is more complicated than the Newtonian description of gravitation, in which there is only a single field. The Newtonian description is a good approximation in most astronomical systems like our solar system because in these systems only one item in the array, g_{00}, has any appreciable variation from point to point in space.

Not only did this theory seem to provide a unification of gravitation and electromagnetism; a subsequent extension of the Kaluza theory due to the physicist Oskar Klein in 1928 seemed to offer hope of solving a problem that had bedeviled Einstein since the advent of General Relativity. In Einstein's 1915 formulation of General Relativity, gravitation appears as a natural consequence of the geometry of space and time, and (aside from possible effects that would vanish at sufficiently large distances) the gravitational field equations are nearly unique. On the other hand, matter is put into the theory "by hand"; there is no a priori way of judging what sorts of matter exist, or how they contribute to the source of the gravitational field. What Klein suggested is that the fifth dimension of Kaluza is not just a formal device, but a real dimension of space, which is curled up so that it is not normally observed, just as a drinking straw may appear one-dimensional when observed casually, but on closer inspection is found to be a two-dimensional sheet, with one dimension curled up.

According to this idea, the fields g_{MN} are no longer arbitrarily constrained to be independent of position in the fifth, curled-up, dimension. Rather, each of them is a sum of an infinite number of terms. Some of these terms *are* independent of the fifth dimension, and these as Kaluza had found represent the four-dimensional gravitational and electromagnetic fields. Then there are new terms that oscillate in the curled-up fifth dimension, like sound waves in an organ pipe, with wavelengths that fit once, twice, or any number of times into the circumference of the curled-up dimension. In four dimensions, these oscillating terms appear like an infinite variety of massive particles, with electric charges proportional to their masses. So not only gravitation and electromagnetism but also charged massive particles arose out of a purely gravitational theory in five dimensions.

The trouble was that particles predicted by the Kaluza-Klein theory could not be electrons or protons or any other known type of elementary particle. Even the lightest of these new particles were predicted to be much too heavy, by about a factor of 10^{19}.

They were so heavy that the gravitational attraction between pairs of these particles would be as strong as their electric attraction or repulsion, which is certainly not the case in ordinary atoms. This setback seems to have caused Einstein to lose interest in theories with extra dimensions, and his efforts in the 1940s turned in a different direction.

In his second main approach to unification, instead of increasing the number of space-time dimensions, Einstein considered the possibility that the array g_{MN} may not be symmetric; that is, g_{MN} might not equal g_{NM}. (In some of his work, Einstein considered the equivalent possibility that the metric has a mathematical property known as hermiticity.) His motivation was an elementary bit of counting. As we have seen, a symmetric 4×4 array has ten independent entries. But without the constraint of symmetry, a 4×4 array has $4^2 = 16$ independent entries. Einstein supposed that the $16 - 10 = 6$ extra fields in a general nonsymmetric metric might represent electromagnetism, which is described by the three components of the electric field and the three components of the magnetic field. It was this hope that preoccupied Einstein in the last decades of his life.

The trouble with this idea is that the ten components of the symmetric part of a general metric and its additional six components have nothing to do with each other. Just putting them together in a 4×4 array says nothing about how their physical properties are related. This is very different from Maxwell's unification of electric and magnetic fields. For one thing, what looks like a purely electric or magnetic field to an observer at rest will look like a combination of electric *and* magnetic fields to a moving observer, while in Einstein's new theory, what looked to one observer like a purely gravitational field would look like a purely gravitational field to all observers. Einstein of course understood this very well, and he kept searching for some physical principle that would tie all sixteen components of the metric together in a natural way, but he never succeeded.

As it has happened, over the half century since his death, Einstein's dream of unification has been partly realized, but in a way very different from what Einstein expected. The theory of electromagnetism is now understood to be part of a larger "electroweak" theory, which describes not just electromagnetism but also certain weak nuclear forces. These forces are responsible for the radioactive processes in atomic nuclei in which neutrons turn into protons, or vice versa. The weak nuclear forces have very short range—the weak force between two nuclear particles drops off steeply at a distance beyond about a quadrillionth of a centimeter; in contrast, electromagnetism, like gravitation, is a long-range force—the force of attraction or repulsion between two charged particles falls smoothly as the inverse square of the distance between them, with no steep drop anywhere. Nevertheless, despite this obvious difference between the electromagnetic and weak nuclear forces, they enter in the same way in the equations of modern electroweak theory, and the difference between them is attributed to properties of the space in which we live, rather than to the theory itself.

Einstein of course knew about radioactivity. It was discovered in 1897, and radioactive salts provided him a decade later with a vivid example of the relation $E = mc^2$ between mass and energy. But as far as I know, Einstein never concerned himself with understanding the weak nuclear forces that make radioactivity possible. Indeed, in his later years Einstein showed no interest in any contemporary work on nuclear and particle physics. Perhaps this was because this work was neither based on nor incorporated General Relativity. Einstein remarked in 1950 that "all attempts to obtain a deeper knowledge of the foundations of physics seem doomed to me unless the basic concepts are in accordance with General Relativity from the beginning." Also, the work of other physicists on nuclear and particle physics was grounded in quantum mechanics, the new probabilistic approach to theoretical physics developed in the 1920s. Einstein regarded quantum mechanics as a renunciation of the traditional aim of physics, to come to a complete understanding of physical reality. Indeed, one of Einstein's hopes for a unified theory was that it would provide an alternative nonquantum mechanical explanation of the atomic phenomena that had already been successfully accounted for by quantum mechanics.

Particle physicists in the 1970s also developed a highly successful theory of one other sort of force, the strong nuclear force that holds quarks together inside the neutron and proton and holds neutrons and protons together inside atomic nuclei. This theory, known as quantum chromodynamics, is mathematically similar to the electroweak theory, so it has not been difficult to imagine various ways to unify both theories in a theory of the electromagnetic, weak and strong nuclear forces.

However it has been much more difficult to bring gravitation into this theoretical framework. The superficial similarity between gravitation and electromagnetism, both producing forces that fall off with the inverse square of the distance, has proved illusory. The one hope for a unification of gravitation with the other forces of nature now lies in the direction of a string theory, a theory that supposes that the fundamental constituents of nature are neither particles nor fields, but strings, one-dimensional entities that are much too small to be seen by us as anything but point particles, but whose various modes of vibration account for the variety of the particle types we observe.

Ironically, these string theories find their most natural formulation in a space-time of ten rather than four dimensions, thus reviving the Kaluza-Klein idea that had so attracted Einstein in the 1930s, though with six extra dimensions rather than one. But of Einstein's other approach to unification, the extension of General Relativity to a nonsymmetric metric, no trace remains in current research. The idea of extra dimensions, especially in the form advanced by Klein, was highly speculative, and may have seemed at first like a purely mathematical game, but right or wrong it had real physical content. In contrast, the idea of a nonsymmetric metric was purely mathematical, and led nowhere. In developing the General Theory of Relativity from 1905 to 1915, Einstein had been guided by an existing mathematical formalism, the Riemann theory of curved space, and perhaps from this experience he had acquired too great a respect for the power of pure mathematics to inspire physical theory. The oracle of mathematics that had served Einstein so well when he was young betrayed him in his later years.

18

Einstein's Mistakes

As a contribution to the World Year of Physics in 2005, *Physics Today* ran a series of articles about Einstein. I was asked to contribute, and sent in the following article, which was published in November 2005.

I chose the subject of Einstein's mistakes because I wanted to illustrate what I take to be one of the strengths of science, that we do not elevate even our greatest heroes into infallible prophets. The set of mistakes discussed in this article is not intended to be exhaustive. Other errors have been dealt with by various commentators: In the March 2005 issue of *Physics Today*, Alex Harvey and Engelbert Schucking had described an erroneous prediction of Einstein regarding the rates of clocks on Earth's surface, and in his 1981 book *Albert Einstein's Special Theory of Relativity*, Arthur I. Miller discussed an error in Einstein's calculation of the electron's transverse mass. The mistakes I discuss here interested me because they revealed something of the intellectual environment in which Einstein worked.

In the following I have added some footnotes to explain a few scientific terms. I have also deleted most of the discussion of Einstein's search for a unified theory of gravitation and electromagnetism that appeared in my *Physics Today* article, because it largely repeated what was said in chapter 17 of this collection.

Albert Einstein was certainly the greatest physicist of the twentieth century, and one of the greatest scientists of all time. It may seem presumptuous to talk of mistakes made by such a towering figure, especially in the centenary of his *annus mirabilis*. But the mistakes made by leading scientists often provide a better insight into the spirit and presuppositions of their times than do their successes. Also, for those of us who have made our share of scientific errors, it is mildly consoling to note that even Einstein made mistakes. Perhaps most important, by showing that we are aware of mistakes made by even the greatest scientists, we set a good example to those who follow other supposed paths to truth. We recognize that our most important scientific forerunners were not prophets whose writings must be studied as infallible guides—they were simply great men and women who prepared the ground for the better understandings we have now achieved.

In thinking of Einstein's mistakes, one immediately recalls what Einstein (in a conversation with George Gamow) called the biggest blunder he had made in his life: the introduction of the cosmological constant. After Einstein had completed the formulation of his theory of space, time, and gravitation—the General Theory of Relativity—he turned in 1917 to a consideration of the space-time structure of the whole universe. He then encountered a problem. Einstein was assuming that, when suitably averaged over many stars, the universe is uniform and essentially static, but the equations of General Relativity did not seem to allow a time-independent solution for a universe with a uniform distribution of matter. So Einstein modified his equations, by including a new term involving a quantity that he called the cosmological constant. Then it was discovered that the universe is not in fact static, but expanding. Einstein came to regret that he had needlessly mutilated his original theory. It may also have bothered him that he had missed predicting the expansion of the universe.

This story involves a tangle of mistakes, but not the one that Einstein thought he had made. First, I don't think that it can count against Einstein that he had assumed the universe is static. With rare exceptions, theorists have to take the world as it is presented to them by observers. The relatively low observed velocities of stars made it almost irresistible in 1917 to suppose that the universe is static. Thus, when Willem de Sitter proposed an alternative solution to the modified Einstein equations in 1917, he took care to use coordinates to identify positions in space and time for which the metric tensor[1] is independent of time. However, the physical meaning of those coordinates is not transparent, and the subsequent realization that de Sitter's alternate cosmology was not in fact static—that matter particles in his model would accelerate away from each other—was at first considered to be a drawback of the theory.

It is true that the American astronomer Vesto Melvin Slipher, while observing the spectra of spiral nebulae in the 1910s, had found a preponderance of redshifts, of the sort that would be produced in an expansion by the Doppler effect,[2] but no one then knew what the spiral nebulae were; it was not until Edwin Hubble found faint Cepheid variables[3] in the Andromeda Nebula in 1923 that it became clear that spiral nebulae were distant galaxies, giant clusters of stars far outside our own galaxy. I don't know if Einstein had heard of Slipher's redshifts by 1917, but in any case he knew very well about at least one other thing that could produce a redshift of spectral lines—a gravitational field—so the mere existence of redshifts would not necessarily convince him of the expansion of the universe.

---

1. Added note: This is the array of fields that characterizes the geometry of space-time. It is discussed in a little more detail in article 17 of this collection.

2. Added note: The Doppler effect is the shift of frequencies of light to lower values, and hence of colors to the red end of the spectrum, that is caused by the motion of a light source away from the observer. (If the source is moving toward the observer then the shift of frequencies is to higher values, and colors shift toward the blue end of the spectrum.)

3. Added note: These are bright variable stars whose period of variation was discovered by Henrietta Swan Leavitt to be tightly correlated with their absolute luminosity, the energy per second of light emitted in all directions. Comparing a Cepheid's apparent luminosity, the energy per second received in each square centimeter of a telescope's mirror, with the absolute luminosity inferred from the Cepheid's period allows a determination of the distance of the star, and hence of the galaxy it inhabits.

It should be acknowledged here that Arthur Eddington, who had learned about General Relativity during World War I from de Sitter, did in 1923 interpret Slipher's redshifts as due to the expansion of the universe in the de Sitter model. Nevertheless, the expansion of the universe was not generally accepted until Hubble announced in 1929—and actually showed in 1931—that the redshifts of distant galaxies increase in proportion to their distance, as would be expected for a uniform expansion. (The faster they go, the farther they get.) Only then was much attention given to the expanding-universe models introduced in 1922 by Alexander Friedmann, based on Einstein's original equations without a cosmological constant. In 1917 it was quite reasonable for Einstein to assume that the universe is static.

Einstein did make a surprisingly trivial mistake in introducing the cosmological constant. Although that step made possible a time-independent solution of the Einstein field equations, the solution described a state of *unstable* equilibrium. The cosmological constant acts like a repulsive force that increases with distance, while the ordinary attractive force of gravitation decreases with distance. Although there is a critical mass density at which this repulsive force just balances the attractive force of gravitation, the balance is unstable; a slight expansion will increase the repulsive force and decrease the attractive force so that the expansion accelerates. It is hard to see how Einstein could have missed this elementary difficulty.

Einstein was also at first confused by an idea he had taken from the philosopher-physicist Ernst Mach: that the phenomenon of inertia[4] is caused by distant masses. To keep inertia finite, Einstein in 1917 supposed that the universe must be finite, and so he assumed that its spatial geometry is that of a three-dimensional spherical surface.[5] It was therefore a surprise to him when calculations showed that if a few particles are introduced into the otherwise

4. Added note: That is, the resistance of masses to being accelerated.
5. Added note: This is the surface of a four-dimensional ball.

empty universe of de Sitter's model, they exhibit all the usual properties of inertia. In General Relativity the masses of distant bodies are not the cause of inertia, though they do affect the choice of inertial frames.[6] But that mistake was harmless. As Einstein pointed out in his 1917 paper, it was the assumption that the universe is static, not that it is finite, that had made a cosmological constant necessary.

Einstein made what from the perspective of today's theoretical physics is a deeper mistake in his dislike of the cosmological constant. In developing General Relativity, he had relied not only on a simple physical principle—the principle of the equivalence of gravitation and inertia that he had developed from 1907 to 1911—but also on a sort of Occam's razor, that the equations of the theory should be not only consistent with this principle but also as simple as possible. In itself, the principle of equivalence would allow field equations of almost unlimited complexity. Einstein could have included terms in the equations involving four space-time derivatives,[7] or six space-time derivatives, or any even number of space-time derivatives, but he limited himself to second-order differential equations, equations in which each term has at most two space-time derivatives.

This could have been defended on practical grounds. The terms in the field equations involving more than two space-time derivatives would have to be accompanied by constant factors proportional to positive powers of some length. If this length was anything like the lengths encountered in elementary particle physics, or even atomic physics, then the effects of these higher derivative terms would be quite negligible at the much larger scales at which all observations

6. Added note: An inertial frame is any coordinate system in which the laws of motion, at least in a small region, are those that apply in Newtonian mechanics (or, more accurately, in the Special Theory of Relativity), in the absence of gravitation. For example, on the surface of the Earth, distances and times measured within a freely falling elevator provide an inertial frame.

7. The derivative of any quantity is its rate of change. In the gravitational field equations, the derivatives are rates of change with shifts in position or time.

of gravitation have so far been made. There is just one modification of Einstein's equations that could have observable effects: the introduction of a term involving no spacetime derivatives at all—that is, a cosmological constant.

But Einstein did not exclude terms with higher derivatives for this or for any other practical reason, but for an aesthetic reason: They were not needed, so why include them? And it was just this aesthetic judgment that led him to regret that he had ever introduced the cosmological constant.

Since Einstein's time, we have learned to distrust this sort of aesthetic criterion. Our experience in elementary particle physics has taught us that any term in the field equations of physics that is allowed by fundamental principles is likely to be there in the equations, though it may be too small at accessible scales of distance and energy to have detectable effects. It is like the ant world in T. H. White's *The Once and Future King:* Everything that is not forbidden is compulsory. Indeed, as far as we have been able to do the calculations, quantum fluctuations by themselves would produce a huge effective cosmological constant, much larger than is allowed by astronomical observations, so that to cancel the unacceptably large contribution of the quantum fluctuations there would have to be a huge "bare" cosmological constant of the opposite sign in the field equations themselves. Occam's razor is a fine tool, but it should be applied to principles, not equations.

It may be that Einstein was influenced by the example of Maxwell's theory, which he had taught himself while a student at the Zurich Polytechnic Institute. James Clerk Maxwell of course invented his equations to account for the known phenomena of electricity and magnetism while preserving the principle of electric charge conservation, and in Maxwell's formulation the field equations contain terms with only a minimum number of space-time derivatives. Today we know that the equations governing electrodynamics contain terms with any number of space-time derivatives, but these terms, like the higher-derivative terms in General Relativity, have no observable consequences at macroscopic scales.

Astronomers in the decades following 1917 occasionally sought signs of a cosmological constant, but they only succeeded in setting an upper bound on the constant. That upper bound was vastly smaller than what would be expected from the contribution of quantum fluctuations, and many physicists and astronomers concluded from this that the constant must be zero. But despite our best efforts, no one could find a satisfactory physical principle that would require a vanishing cosmological constant.

Then in 1998, measurements of redshifts and distances of supernovae by the Supernova Cosmology Project and the High-z Supernova Search Team showed that the expansion of the universe is accelerating, as de Sitter had found in his model. It seems that about 70 percent of the energy density of the universe is a sort of "dark energy," filling all space. This was subsequently confirmed by observations of the angular size of anisotropies in the cosmic microwave background.[8] The density of the dark energy is not varying rapidly as the universe expands, and if it is truly time-independent, then it is just the effect that would be expected from a cosmological constant. However this works out, it is still puzzling why the cosmological constant is not as large as would be expected from calculations of quantum fluctuations. In recent years the question has become a major preoccupation of theoretical physicists. Regarding his introduction of the cosmological constant in 1917, Einstein's real mistake was that he thought it was a mistake.

A historian, reading the foregoing in a first draft of this article, commented that I might be accused of perpetrating Whig history.

---

8. Added note: The cosmic microwave background is radiation left over from the time when the universe first became transparent to radiation, about four hundred thousand years after the start of the present phase of the expansion of the universe. It has been cooled by the expansion of the universe from a temperature at that time of 3,000 degrees Kelvin to a present average temperature of 2.725 degrees Kelvin (that is, Celsius degrees above absolute zero). But this temperature varies from point in the sky by amounts of about one part in one hundred thousand, and by studying these variations it is possible to infer that the expansion of the universe since this radiation was emitted has been accelerated, presumably by dark energy.

The term "Whig history" was coined in a 1931 lecture by the historian Herbert Butterfield. According to Butterfield, Whig historians believe that there is an unfolding logic in history, and they judge the past by the standards of the present. But it seems to me that, although Whiggery is to be avoided in political and social history (which is what concerned Butterfield), it has a certain value in the history of science. Our work in science is cumulative. We really do know more than our predecessors, and we can learn about the things that were not understood in their times by looking at the mistakes they made.

The other mistake that is widely attributed to Einstein is that he was on the wrong side in his famous debate with Niels Bohr over quantum mechanics, starting at the Solvay Congress of 1927 and continuing into the 1930s. In brief, Bohr had presided over the formulation of a "Copenhagen interpretation" of quantum mechanics, according to which it is only possible to calculate the probabilities of the various possible outcomes of experiments. Einstein rejected the notion that the laws of physics could deal with probabilities, famously decreeing that God does not play dice with the cosmos. But history gave its verdict against Einstein—quantum mechanics went on from success to success, leaving Einstein on the sidelines.

All this familiar story is true, but it leaves out an irony. Bohr's version of quantum mechanics was deeply flawed, but not for the reason Einstein thought. The Copenhagen interpretation describes what happens when an observer makes a measurement, but the observer and the act of measurement are themselves treated according to the rules of prequantum classical physics. This is surely wrong: Physicists and their apparatus must be governed by the same quantum mechanical rules that govern everything else in the universe. But these rules are expressed in terms of a wave function (or, more precisely, a state vector) that evolves in a perfectly deterministic way. So where do the probabilistic rules of the Copenhagen interpretation come from?

Considerable progress has been made in recent years toward the resolution of the problem, which I cannot go into here. It is enough to say that neither Bohr nor Einstein had focused on the real problem with quantum mechanics. The Copenhagen rules clearly work, so they have to be accepted. But this leaves the task of explaining them by applying the deterministic equation for the evolution of the wave function, the Schrödinger equation, to observers and their apparatus. The difficulty is not that quantum mechanics is probabilistic—that is something we apparently just have to live with. The real difficulty is that it is also deterministic, or more precisely, that it combines a probabilistic interpretation with deterministic dynamics.

Einstein's rejection of quantum mechanics contributed, in the years from the 1930s to his death in 1955, to his isolation from other research in physics, but there was another factor. Perhaps Einstein's greatest mistake was that he became the prisoner of his own successes. It is the most natural thing in the world, when one has scored great victories in the past, to try to go on to further victories by repeating the tactics that previously worked so well. Think of the advice given to Egypt's President Gamal Abdel Nasser by an apocryphal Soviet military attaché at the time of the 1956 Suez crisis: "Withdraw your troops to the center of the country, and wait for winter."

And what physicist had scored greater victories than Einstein? After his tremendous success in finding an explanation of gravitation in the geometry of space and time, it was natural that he should try to bring other forces along with gravitation into a "unified field theory" based on geometrical principles. Since electromagnetism was the only other force that in its macroscopic effects seemed to bear any resemblance to gravitation, it was the hope of a unification of gravitation and electromagnetism that drove Einstein in his later years. But his efforts all failed.

Even though it was a mistake for Einstein to turn away from the exciting progress being made in the 1930s and 1940s by younger physicists, it revealed one admirable feature of his personality. Einstein never wanted to be a mandarin. He never tried to induce physicists in general to give up their work on nuclear and particle physics and follow his ideas. He never tried to fill professorships at the Institute for Advanced Studies with his collaborators or acolytes. Einstein was not only a great man, but a good one. His moral sense guided him in other matters: He opposed militarism during World War I; he refused to support the Soviet Union in the Stalin years; he became an enthusiastic Zionist; he gave up his earlier pacifism when Europe was threatened by Nazi Germany, for instance urging the Belgians to rearm; and he publicly opposed McCarthyism. About these great public issues, Einstein made no mistakes.

19

Living in the Multiverse

There has been a good deal of interest lately in the possibility that what we call our universe, the expanding cloud of galaxies that we see extending for at least ten billion light years in all directions, is only an episode in a much larger "multiverse," composed of many subuniverses, some like our own but most very different. I was invited to give a talk on this subject as the opening talk at a symposium, "Expectations of a Final Theory," to be held at Trinity College, Cambridge, in September 2005. The setting was superb—it was upstairs at the Master's Lodge, with a large oil portrait of a former fellow of the college, Isaac Newton, looking dourly down at the speakers. Though as usual I spoke from an outline, I wrote up the talk afterwards for inclusion in a book edited by Bernard Carr, *Universe or Multiverse?* (Cambridge University Press, 2007).

The middle part of my talk was quite technical, so I have cut most of it out of the version presented here. Also, this part of my talk included many references to physicists' names and scientific articles (including my own), which I have dropped in the present version. But in order to keep the sense of the anecdotes at the end of the article, I have had to leave in the names of three scientists. Their names will be familiar to physicists and astrophysicists, but may not be known to some general readers. Suffice it to say that David Gross is a theoretical physicist, currently the director of the Kavli Institute for Theoretical Physics at Santa Barbara; Andrei Linde is an astrophysicist at Stanford; and Martin Rees (a.k.a. Lord Rees) is an astrophysicist at Cambridge, and as Master of Trinity College was the host of our meeting.

We usually mark advances in the history of science by what we learn about nature, but at certain critical moments the most important thing is what we discover about science itself. These discoveries lead to changes in how we score our work, in what we consider to be an acceptable theory.

For an example look back to a discovery made just one hundred years ago. Before 1905 there had been numerous unsuccessful efforts to detect changes in the speed of light due to the motion of the earth through the ether. Attempts were made by Fitzgerald, Lorentz, and others to construct a mathematical model of the electron (which was then conceived to be the chief constituent of all matter) that would explain how rulers contract when moving through the ether in just the right way to keep the apparent speed of light unchanged. Einstein instead offered a principle of invariance, which stated that not just the speed of light but all the laws of nature are unaffected by a transformation to a frame of reference in uniform motion. Lorentz grumbled that Einstein was simply assuming what he and others had been trying to prove. But history was on Einstein's side. The 1905 Special Theory of Relativity was the beginning of a general acceptance of principles of invariance as a valid basis for physical theories.

This was how Special Relativity made a change in science itself. From one point of view, Special Relativity was no big thing—it just amounted to the replacement of the principle of invariance obeyed by Newtonian mechanics, invariance under translations, rotations, and uniform additions to velocity, known as Galilean invariance, with a different principle of invariance, invariance under translations, rotations, and what are called Lorentz transformations, all of which leave the speed of light unchanged. But never before had a principle of invariance been taken as a legitimate hypothesis on which to base a physical theory.

As usually happens with this sort of revolution, Einstein's advance came with a retreat in another direction: The effort to construct a mechanical model of the electron was permanently abandoned. Instead, principles of invariance increasingly became the dominant foundation for physical theories. This tendency was accelerated after the advent of quantum mechanics in the 1920s, because the survival of invariance principles in quantum theories imposes highly restrictive consistency conditions (including the existence of antimatter) on physically acceptable theories. Our present Standard Model of elementary particle interactions can almost be regarded as the consequence of certain principles of invariance and the associated quantum mechanical consistency conditions.

The development of the Standard Model did not involve any changes in our conception of what was acceptable as a basis for physical theories. Indeed, the Standard Model can be regarded as just the quantum electrodynamics of the 1940s, writ large. Similarly, when the effort to extend the Standard Model to include gravity led to widespread interest in string theory, we expected to score the success or failure of this theory in the same way as for the Standard Model: String theory would be a success if its invariance principles and consistency conditions led to a successful prediction of the particle masses and other parameters of the Standard Model.

Now we may be at a new turning point, a radical change in what we accept as a legitimate foundation for a physical theory. The current excitement is of course a consequence of the recent discovery of a vast number of solutions of string theory. The consistency conditions of string theory are most naturally satisfied in ten space-time dimensions. In order to account for the fact that we only perceive one time and three space dimensions, it is necessary to suppose that the extra six dimensions are tightly rolled up. The shape of this compact six-dimensional space is characterized by a large number of integers, which specify the way that various fields thread through holes and handles in the space, and each of these integers can take a large number of values. It has been estimated that the number of solutions for the way that the six extra dimensions are rolled up is of order 10^{100} to 10^{500}—that is, a one followed by one hundred to five hundred zeroes. String theorists have picked up the term "string landscape" for this multiplicity of solutions from biochemistry, where the possible choices of orientation of each chemical bond in large molecules leads to a vast number of possible configurations. Unless one can find a reason to reject all but a few of the string theory solutions, we may have to accept that much of what we had hoped to calculate are environmental parameters, characterizing the particular solution of the equations of string theory that describes our subuniverse. Then these constants would be like the distance of the Earth from the Sun, whose value we will never be able to deduce from first principles.

We lose some, and win some. The larger the number of possible values of physical parameters provided by the string landscape, the more string theory legitimates anthropic reasoning as a new basis for physical theories: Any scientists who study nature must live in a part of the landscape where physical parameters take values suitable for the appearance of life and its evolution into scientists, and it is likely that such parts of the landscape exist if physical parameters take sufficiently many different values in different parts of the string landscape.

An apparently successful example of anthropic reasoning was already at hand by the time the string landscape was discovered. For decades there seemed to be something peculiar about the value of the vacuum energy, the energy of empty space, apart from any matter or radiation it may contain. [This, also known as dark energy, is the subject of article 5 in this collection.] Quantum fluctuations in known fields at well-understood energies (say, less than one hundred times the energy in a proton mass) give a value of the vacuum energy larger than observationally allowed by a factor 10^{56}. This contribution to the vacuum energy might be cancelled by quantum fluctuations of higher energy, or by simply including a suitable "cosmological constant" term in the Einstein field equations, but the cancellation would have to be exact to fifty-six decimal places. No invariance argument or adjustment mechanism could be found that would explain such a cancellation. Even if such an explanation could be found, there would be no reason to suppose that the remaining net vacuum energy would be comparable to the *present* value of the matter density, and since it is certainly not very much larger, it was natural to suppose that it is very much less, too small to be detected.

On the other hand, if the vacuum energy takes a broad range of values in the multiverse, then it is natural for scientists to find themselves in a subuniverse in which the vacuum energy takes a value suitable for the appearance of scientists. It was pointed out in 1987 that this value for the vacuum energy can't be too large and positive, because then galaxies and stars would not form. Roughly, the limit is that the vacuum energy should be less than the mass density of the universe at the time when galaxies first condense. Since this was in the past, when the mass density of the expanding universe was larger than at present, the anthropic upper limit on the vacuum energy density is larger than the present mass density, but not many orders of magnitude greater.

But anthropic arguments provide not just a bound on the vacuum energy; they give us some rough idea of the value to be expected: The vacuum energy should be not very different from the average of the values suitable for life. This is sometimes called the "principle of mediocrity." This average is positive, because if the vacuum energy were negative, it would have to be less in absolute value than the mass density of the universe during the whole time that life evolves, since otherwise the universe would collapse before any astronomers come on the scene, while if positive, the vacuum energy only has to be less than the mass density of the universe at the time when most galaxies form, giving a much broader range of possible positive than negative values. In 1997–98 a detailed calculation of the probability distribution of values of the vacuum energy that would be seen by astronomers throughout the multiverse (under the assumption that the a priori probability distribution is flat in the relatively very narrow range that is anthropically allowed) showed that if vacuum energy turned out to make up less than about 60 percent of the total energy of the present universe, anthropic reasoning could not explain why it was so small.

As it happens, it turned out that the vacuum energy density is not too small. In 1998, observations of type Ia supernovae revealed that the expansion of our universe is accelerating, and drew the conclusion that vacuum energy makes up about 70 percent of the energy of our present universe. This is still a bit less than what one might have expected. Using the best modern data about the fluctuations that grow into galaxies, the probability of a random astronomer anywhere in the multiverse seeing a vacuum energy this small or smaller is calculated to be only about 16 percent. But at least it would not be amazing to find ourselves in a part of the multiverse with a vacuum energy like the one we observe.

Now I want to take up four problems we have to face in working out the anthropic implications of the string landscape.

1. What is the shape of the string landscape?

We need to find the statistical rules governing different string solutions: how many solutions there are with physical parameters taking values in various different ranges. This is made difficult because we still don't really know the fundamental principles underlying string theory.

2. What constants scan?

To allow the application of anthropic reasoning to a particular constant of nature, the variation of this constant from one part of the multiverse to another must be sufficiently broad and smooth. More specifically, anthropic reasoning makes sense for a given constant if the range over which the constant varies in the landscape is large compared with the anthropically allowed range of values of the constant, for then it is reasonable to assume that the a priori probability distribution is flat in the narrow anthropically allowed range. We need to know what constants actually "scan" in this sense. Physicists would like to be able to calculate as much as possible, so we hope that not too many constants scan, or vary at all from one subuniverse to another. The most optimistic hypothesis is that the only constants that vary are the vacuum energy, and perhaps the overall mass scale of observed elementary particles.

3. How should we calculate anthropically conditioned probabilities?

We would expect the probability that a random scientific civilization would observe the value of a given physical constant to be in a given range to be proportional to the number of scientific civilizations in the part of the string landscape for which the constant is in that range. In the calculations mentioned above, this number was taken to be proportional to the *fraction* of nuclear particles that find themselves in galaxies, whose stars have planets that can support life, but what if the total number of these particles itself scans? What if it is infinite?

4. How is the landscape populated?

There are at least four ways in which we might imagine the different "universes" described by the string landscape actually to exist: The various subuniverses may be simply different regions of space, or they may be different eras of time in a single big bang, or they may be disconnected regions of space-time, or they may be different parts of the mathematical space known as Hilbert space inhabited by the wave function of the universe. These alternatives are by no means mutually exclusive. In particular, it seems to me that, whatever one concludes about the first three alternatives, we will still have the possibility that the wave function of the universe is a superposition of different terms, representing different ways of populating the landscape in space and/or time.

In closing, I would like to comment about the impact of anthropic reasoning within and beyond the physics community. Some physicists have expressed a strong distaste for anthropic arguments. (I have heard David Gross say "I hate it.") This is understandable. Theories based on anthropic calculation certainly represent a retreat from what we had hoped for: the calculation of all fundamental parameters from first principles. It is too soon to give up on this hope, but without loving it we may just have to resign ourselves to a retreat, just as Newton had to give up Kepler's hope of a calculation of the relative sizes of planetary orbits from first principles.

There is also a less well-founded reason for hostility to the idea of a multiverse, based on the fact that we will never be able to observe any subuniverses except our own. There are various other ingredients of accepted theories that we will never be able to observe, without our being led to reject these theories. The test of a physical theory is not that everything in it should be observable and every prediction it makes should be testable, but rather that enough is observable and enough predictions are testable to give us confidence that the theory is right.

Finally, I have heard the objection that, in trying to explain why the laws of nature are so well suited for the appearance and evolution of life, anthropic arguments take on some of the flavor of religion. I think that just the opposite is the case. Just as Darwin and Wallace explained how the wonderful adaptations of living forms could arise without supernatural intervention, so the string landscape may explain how the constants of nature that we observe can take values suitable for life without being fine-tuned by a benevolent creator. I found this parallel well understood in a surprising place, a *New York Times* op-ed article in July 2005 by Christoph Schönborn, Cardinal Archbishop of Vienna. His article concludes as follows:

> Now, at the beginning of the 21st century, faced with scientific claims like neo-Darwinism and the multiverse hypothesis in cosmology invented to avoid the overwhelming evidence for purpose and design found in modern science, the Catholic Church will again defend human nature by proclaiming that the imma-

nent design evident in nature is real. Scientific theories that try to explain away the appearance of design as the result of "chance and necessity" are not scientific at all, but, as John Paul put it, an abdication of human intelligence.

It's nice to see work in cosmology get some of the attention given these days to evolution, but of course it is not religious preconceptions like these that can decide any issues in science.

It must be acknowledged that there is a big difference in the degree of confidence we can have in neo-Darwinism and in the multiverse. It is settled, as well as anything in science is ever settled, that the adaptations of living things on earth have come into being through natural selection acting on random undirected inheritable variations. About the multiverse, it is appropriate to keep an open mind, and opinions among scientists differ widely. In the Austin airport on the way to this meeting I noticed for sale the October issue of a magazine called *Astronomy,* having on the cover the headline "Why You Live in Multiple Universes." Inside I found a report of a discussion at a conference at Stanford, at which Martin Rees said that he was sufficiently confident about the multiverse to bet his dog's life on it, while Andrei Linde said he would bet his own life. As for me, I have just enough confidence about the multiverse to bet the lives of both Andrei Linde *and* Martin Rees's dog.

20

Against the Boycott

In May 2006 I learned that in Great Britain the National Association of Teachers in Further and Higher Education (Natfhe) was considering a motion calling on its members to engage in private boycotts of Israeli academic institutions and professors. I joined with a group of Nobel laureates[1] in opposing this motion, and the *Times Higher Education Supplement* of London invited me to submit a statement of my views. The following statement was published in their May 26, 2006 issue, along with an edited version of an open letter supporting the boycott, issued by the Federation of Unions of Palestinian University Professors and Employees and the Palestinian Campaign for the Academic and Cultural Boycott of Israel.

If the lecturers' union Natfhe votes to boycott Israeli academics who refuse to oppose Israel's policies, then it will deserve the moral condemnation of the world. Israel is a democracy that extends full civil rights to all citizens—Arabs as well as Jews. It is in the course of withdrawing from Gaza and most of the West Bank, and it actively pursues ties with Arab academic institutions.

If the urge to boycott is irresistible, why not boycott academics in Sudan, where a government-supported militia rapes and murders blacks? Why not boycott academics in Saudi Arabia, where no Jew or Christian is allowed to become a citizen? Why not boy-

---

1. Aaron Ciechanover, David Gross, Dudley Herschbach, David Kahneman, Walter Kohn, Jean-Marie Lehn, and Frank Wilczek.

cott academics in Iran, where courts throw Jews into jail on trumped-up espionage charges? For that matter, why not boycott academics in all countries that have adopted Islamic law (sharia), which discriminates against women and makes it a capital offense for Muslims to renounce Islam?

Perhaps one could look beyond the issue of discrimination and boycott academics in North Korea, which has the most repressive government on Earth, or those in Gaza, where a government of terrorists has just taken power.

Yet for the past decade it is only Israel that British academic unions have considered boycotting. I can attribute this only to a really spectacular moral blindness, a hatred of Jews, or both.

It is never a good idea for academics to boycott colleagues in other countries on political grounds. During the Cold War, American and Soviet scientists were careful to keep intellectual communication open; this not only served the cause of science, but promoted personal relationships that led to initiatives in arms control. In a similar spirit, when I ran the Jerusalem Winter School of Theoretical Physics, we did what we could to recruit Arab students from Muslim countries whose governments discriminated against Jews. We never dreamt of boycotting them.

The Natfhe draft proposal blames Israel for "construction of the exclusion wall." This barrier does impose a nuisance on both Arabs and Jews. However, it is not being built because Jews do not want to associate with Arabs, but because they do not want to be murdered by them.

There was no thought of a wall until the intifada in 2002 reached new heights of brutality. In fact, the Israeli wall is not very different from the thirteen-mile "Peace Line" in Belfast, built to curb violence between Catholic and Protestant neighborhoods. Even though incomplete, the barrier works: it has already greatly reduced the number of deaths of Israelis at the hands of suicide bombers. Since the wall saves lives, one marvels at the callousness of a call for Israelis to die so that Arabs will not be inconvenienced.

The other sin that the Natfhe draft proposal lays at the door of Israel is "discriminatory educational practices." I cannot conceive what is meant by this. Israel's universities are open to its Arab citizens on the same terms as its Jewish citizens, and Arab students attend the secular universities of Haifa, Tel Aviv, and Jerusalem in large numbers. Of course, noncitizens are treated differently. The state universities of Texas naturally give preferential treatment in admissions and tuition charges to citizens of Texas. But no one has proposed to boycott Texas academics on the grounds of discrimination against Oklahomans.

Israel is the only true democracy in the Levant, and it is the only country in the world whose very existence is threatened by its neighbors. Let no one think that the issue is the Israeli presence in the West Bank or Gaza. The issue is the Jewish presence and, even worse, Jewish success, in the midst of Islam. Jews were massacred by Arabs in Palestine before there was an Israel, and Israel was attacked again and again when the West Bank and Gaza were entirely in Arab hands. Saudi Arabia and Syria consider themselves still at war with Israel, while Hamas and the President of Iran openly call for Israel's annihilation. Far from meriting a boycott, Israel deserves the support of academics, especially in a country such as Great Britain that did so much to preserve democracy in two world wars.

If Natfhe does call for a boycott of Israeli academics, then I would have one request of its members. It is to include me among those boycotted. I would consider it an honor.

———

This was not the end of my involvement with the boycott issue. The boycott motion passed at the 2006 Natfhe meeting, and with regret I felt I had to withdraw my acceptance of an invitation to speak at a conference on cosmology at the University of Durham that summer. I did not want to start a counter-boycott of Britain— I just felt that as a liberal, as an academic, and as a Jew, I had to make some personal gesture of protest. I was glad to learn later that the boycott was lifted, though by then it was too late for me to attend the conference at Durham.

Then a few months later I received an invitation to speak at Imperial College, London, from an old friend, Professor Michael Duff. I was asked to give a talk in July 2007 in honor of the late Abdus Salam, with whom I had a fruitful collaboration during a stay at Imperial College in the 1961–62 academic year, and also to give a technical talk at an international physics conference to be held in London that July. Since the Natfhe boycott had been lifted, and I was glad to have a chance to honor Salam, I accepted. Alas, in May 2007 I learned that the National Union of Journalists of Great Britain had voted at its national conference to boycott Israeli products. Similar initiatives were under consideration in Britain by the University and College Union and the Universities of Brighton and of East London, and 180 British doctors were calling for a boycott of the Israeli Medical Association. I wrote to Duff to tell him that I would not be coming to London in July. In my letter I explained "I know that some will say that these boycotts are directed only against Israel, rather than generally against Jews. But given the history of attacks on Israel and the oppressiveness and aggressiveness of other countries in the Middle East and elsewhere, boycotting Israel indicates a moral blindness for which it is hard to find any other explanation than anti-Semitism. (The only other explanation I can imagine is a desire to pander to the growing Moslem minority in Britain.) . . . I regret that I will not have a chance to speak in praise of Salam, who I like to think would have been sorry for this action of British journalists."

I told Duff that he should feel free to show my letter to anyone interested. Somehow it reached the press in Britain. On May 24 the *Guardian* newspaper published excerpts from my letter to Duff, and on May 25 the *Times Higher Education Supplement* printed the whole letter, and quoted a telephone interview with me, as follows: "I love England. I was married there, and am a foreign member of the Royal Society, of which I am very proud. Undoubtedly I will return, but at this moment I feel I have to make some gesture of moral protest. There are so many countries that deserve condemnation and yet are given a free pass. I can think of no other explanation for Israel's treatment than anti-Semitism. This attitude is a feature of the left-wing intelligentsia in England, which is sad for me as I consider myself a man of the Left."

In 2007 the University and College Union (which had absorbed Natfhe) passed a motion calling for a boycott of Israeli academics, which was then overturned by a special conference. In 2008 a conference of this Union called on members to "consider the moral and political implications of educational links with Israeli institutions," but it seems subsequently to have abandoned attempts to implement this motion. I have not returned to England since this all started, but doubtless eventually I will.

21

A Deadly Certitude

This is a review of *The God Delusion,* by Richard Dawkins (Bantam, London, 2006). I don't review many books, but I agreed when the *Times Literary Supplement* of London asked me to review this one, because I had disagreed with criticisms of the book in the *New Republic* and the *London Review of Books,* and because I wanted to weigh in with some views of my own on the issues covered in Dawkins's book. This review appeared in the *TLS* in January 2007.

Of all the scientific discoveries that have disturbed the religious mind, none has had the impact of Darwin's theory of evolution by natural selection. No advance of physics or even cosmology has produced such a shock.

In the early days of Christianity the Church Fathers Theophilus of Antioch and Lactantius rejected the knowledge, common since the time of Plato, that the Earth is a sphere. They insisted on the literal truth of the Bible, and from Genesis to Revelations verses could be interpreted to mean that the Earth is flat. But the evidence for a spherical Earth was overwhelming to anyone who had seen a ship's hull disappear below the horizon while its masts were still visible, and in the end the flat Earth did not seem worth a fight.

The more radical idea that the Earth moves around the Sun was harder to accept. After all, the Bible puts mankind at the center of a great cosmic drama of sin and salvation, so how could our Earth be just another planet? Until the nineteenth century Copernican astronomy could not be taught at Salamanca or other Spanish universities, but by Darwin's time it troubled hardly anyone.

A different challenge to religion emerged with Newton. His theories of motion and gravitation showed how natural phenomena could be explained without divine intervention (though Newton himself remained deeply religious), and were opposed on religious grounds at Newton's own university by John Hutchinson. But opposition to Newtonianism in Europe collapsed before the close of the eighteenth century. Believers could comfort themselves with the thought that miracles were simply occasional exceptions to Newton's laws, and anyway mathematical physics was unlikely to disturb those who did not understand its explanatory power.

Darwinism was different. It was not just that the theory of evolution, like the theory of a spherical moving Earth, is in conflict with biblical literalism; it was not just that evolution, like the Copernican theory, denied a central status to humans; and it was not just that evolution, like Newton's theory, provided a nonreligious explanation for natural phenomena that had seemed inexplicable without divine intervention. Much worse, among the natural phenomena explained by natural selection were the very features of humanity of which we are most proud. It became plausible that our love for our mates and children, and, according to the work of modern evolutionary biologists, even more abstract moral principles such as loyalty, charity, and honesty, have an origin in evolution, rather than in a divinely created soul.

Given the battering that traditional religion has taken from the theory of evolution, it is fitting that the most energetic, eloquent, and uncompromising modern adversaries of religion are biologists who helped us understand evolution: first Francis Crick, and now Richard Dawkins. In his best-seller *The God Delusion,* Dawkins caps a series of his books on biology and religion with a swingeing attack on every aspect of religion—not just traditional religion but also the vaguer modern assortment of pieties that often appropriates the name of religion.

In the unkindest cut of all, Dawkins even argues that the persistence of belief in God is itself an outcome of natural selection—acting perhaps on our genes, as argued by Dean Hamer in *The God Gene,* but more certainly on our *memes,* the bundles of cultural beliefs and attitudes that in a Darwinian though nonbiological way tend to be passed on from generation to generation. It is not that the meme helps the believer or the believer's genes to survive; it is the meme itself that by its nature tends to survive.

For instance, the persistence of belief in a particular religion is naturally aided if that religion teaches that God punishes disbelief. Such a religion will tend to survive if the threatened punishment is sufficiently awful. In contrast, a religion would have trouble keeping converts in line if it taught that infidels are subject after death to only a brief spell of mild discomfort, after which they join the faithful in eternal bliss. So it is natural that in traditional Christianity and Islam, disbelief became the ultimate crime, and hell the ultimate torture chamber. No wonder the mathematician Paul Erdos habitually referred to God as the Supreme Fascist.

Incidentally, Dawkins's book (like this review) focuses on Christianity and Islam, which traditionally emphasize the importance of belief, rather than on religions like Judaism or Hinduism or Shinto, which are tied to specific ethnic groups, and stress observance rather than faith.

Dawkins, like Erdos, dislikes God. He calls the God of the Old Testament "the most unpleasant character in all fiction: jealous and proud of it; a petty, unjust unforgiving control freak; a vindictive, bloodthirsty ethnic cleanser; a misogynist, pestilential, megalomaniacal, sadomasochistic, capriciously malevolent bully." As to the New Testament, he quotes with approval the opinion of Thomas Jefferson, that "The Christian God is a being of a terrific character—cruel, vindictive, capricious, and unjust."

This is strong stuff, and Dawkins obviously intends to shock the reader, but this invective has a constructive purpose. By attacking the God of sacred scripture, he is trying to weaken the authority of that God's supposed commands, commands whose interpretation has led humanity to a shameful history of inquisitions, crusades, and jihads. Dawkins treats the reader to many brutal details, but the reader only has to look at today's headlines to supply his or her own. For some reason, Dawkins does not comment on the God of the Koran, who would seem to provide equal opportunities for invective.

The reviews of *The God Delusion* in the *New York Times* and the *New Republic* took Dawkins to task for his contemptuous rejection of the classic "proofs" of the existence of God. I agree with Dawkins in his rejection of these proofs, but I would have answered them a little differently.

The "ontological proof" of St. Anselm asks us first to agree that it is possible to conceive of something than which nothing greater can be conceived. Once that agreement is obtained, the sly philosopher points out that the thing conceived of must exist, since if it did not, then something just like it that actually exists would thereby be greater. And what could be this greatest actually existing thing, but God? Q.E.D. From the monk Gaunilo in Anselm's time to philosophers in our own such as J. L. Mackie and Alvin Plantinga, there is general agreement that Anselm's proof is flawed, although they disagree about what the flaw is. My own view is that the proof is circular: it is not true that one can conceive of something than which nothing greater can be conceived unless one first assumes the existence of God.

Anselm's "proof" has reappeared in many other forms, from Descartes to Plantinga, and in these various forms it has been refuted by philosophers from Kant and Hume to Mackie. It is a little like an infectious disease that can be defeated by an antibiotic, but which then evolves so that it needs to be defeated all over again.

The "cosmological proof" is no better logically, but it does have a certain appeal to the physicist. In essence, it argues that everything has a cause, and since this chain of causality can not go on forever, it must terminate in a first cause, which we call God. The idea of an ultimate cause is deeply attractive, and indeed the dream of elementary particle physics is to find the final theory that we think lies at the root of all chains of explanation of what we see in nature. The trouble is that such a mathematical final theory would hardly be what anyone means by God. Who prays to quantum mechanics?

The believer may justly argue that no theory of physics can be a first cause, since we would still wonder why nature is governed by that theory, rather than some other. Yet in just the same sense, God cannot be a first cause either, for whatever our conception of God we could still wonder why the world is governed by that sort of God, rather than some other.

The "proof" that has historically been most persuasive is the argument from design. The world in general (and life in particular) is supposed to be so marvelously shaped that it could only have been the handiwork of the supreme Designer. The great achievement of the work of scientists from Newton to Crick and Dawkins has been to refute this argument by explaining the world.

I find it disturbing that Thomas Nagel in the *New Republic* dismisses Dawkins as an "amateur philosopher," while Terry Eagleton in the *London Review of Books* sneers at Dawkins for his lack of theological training. Are we to conclude that opinions on matters of philosophy or religion are only to be expressed by experts, not mere scientists or other common folk? It is like saying that only political scientists are justified in expressing views on politics. Eagleton's judgment is particularly inappropriate; it is like saying that no one is entitled to judge the validity of astrology who cannot cast a horoscope.

Where I think Dawkins goes wrong is that, like Henry the Fifth after Agincourt, he does not seem to realize the extent to which his side has won. Setting aside the rise of Islam in Europe, the decline of serious Christian belief among Europeans is so widely advertised that Dawkins turns to America for most of his examples of unregenerate religious belief. He attributes the greater regard for religion in America to the fact that we have never had a national established church, an idea he may have picked up from Toqueville. But although most Americans may be sure of the value of religion, as far as I can tell they are not very certain about the truth of what their own religion teaches. According to a recent article in the *New York Times*, American evangelists are in despair over a poll that showed that only 4 percent of American teenagers will be "Bible-believing Christians" as adults.

The spread of religious toleration provides evidence of the weakening of religious certitude. Most Christian sects have historically taught that there is no salvation without faith in Christ. If you are really sure that anyone without such faith is doomed to an eternity of hell, then propagating that faith and suppressing disbelief would logically be the most important thing in the world—far more important than any merely secular virtues like religious toleration. Yet religious toleration is rampant in America. No one in America who publicly expressed disrespect for any particular religion could be elected to a major office.

Even though American atheists might have trouble winning elections, Americans are fairly tolerant of us unbelievers. My many good friends in Texas who are professed Christians do not even try to convert me. This might be taken as evidence that they don't really mind if I spend eternity in hell, but I prefer to think (and Baptists and Presbyterians have admitted it to me) that in fact they are not all that certain about hell and heaven.

I have often heard the remark that it is not so important what one believes; the important thing is how we treat each other. This of course is good to hear. But "not important what one believes"? Imagine trying to explain that to Luther, or Calvin, or St. Paul! Remarks like this show a massive retreat of Christianity from the ground it once occupied, a retreat that can be attributed to no new revelation, but only to a loss of certitude.

Much of the weakening of religious certitude in the Christian West can be laid at the door of science. Even people whose religion might incline them to hostility to the pretensions of science generally understand that they have to rely on science rather than religion to get things done. I encountered an example of this in a drive north from Austin to Waco a few years ago. It was a time when creationists were arguing that scientific theories like Darwinian evolution were themselves a form of religion, a religion known as "secular humanism," which the public school curriculum should not favor over other religions, like Christianity. I was on my way to Baylor University to participate in a conference on this issue. This was also a time of drought in Texas, and as I drove through the Bible Belt farming country of the Brazos valley I thought of all the farmers who must be praying for rain. Then I turned on the radio, and heard a weather forecast from a local station. The announcer spoke of temperature and humidity, and of cold fronts coming in from West Texas, but there was not one word about all the praying that must be going on. It occurred to me to wonder, why would Bible Belt listeners put up with this secular humanist weather forecast? Obviously, it was because they really wanted to know if it was going to rain.

The role of science in weakening religious certitude is one of its greatest contributions to civilization. But this has not happened to anything like the same extent in the world of Islam. One finds in Islamic countries not only religious opposition to specific scientific theories, as occasionally in the West, but a widespread religious hostility to science itself. My late friend, the distinguished Pakistani physicist Abdus Salam, tried to convince the rulers of the oil-rich states of the Persian Gulf to invest in scientific education and research, but he found that although they were enthusiastic about technology, they felt that pure science presented too great a challenge to faith. The Muslim Brotherhood of Egypt at one point called for an end to scientific education. This is despite the fact that in the ninth century, when science barely existed in Europe, the greatest center of scientific research in the world was the House of Wisdom in Baghdad.

Islam began to turn against science in the twelfth century. The most influential figure was the philosopher Abu Hamid al-Ghazali, who argued in *The Incoherence of the Philosophers* against the very idea of laws of nature, on the ground that any such laws would put God's hands in chains. According to al-Ghazali, a piece of cotton placed in a flame does not darken and smolder because of the heat, but because God wants it to darken and smolder. In the centuries after al-Ghazali, science went into a decline in Islamic countries that has not yet been reversed.

The consequences are hideous. Whatever one thinks of the Muslims who blow themselves up in crowded cities in Europe or Israel or fly planes into buildings in the United States, who could dispute that the certitude of their faith has something to do with it?

George Bush and many others would have us believe that terrorism is a distortion of Islam, and that Islam is a religion of peace. Of course, it's good policy to say this, but statements about what "Islam is" make little sense. Islam like all other religions was created by people, and there are potentially as many different versions of Islam as there are people who profess to be Muslims. (The same remarks apply to Eagleton's highly personal account of what Christianity "is.") I don't know on what ground one can say that a peaceable well-intentioned person like Abdus Salam was any more a true Muslim than the murderous holy warriors of Hezbollah and Islamic Jihad, or the clerics throughout the world of Islam who incite hatred and violence, or those Muslims who demonstrate against supposed insults to their faith but not against the atrocities committed in its name. (Incidentally, Abdus Salam regarded himself as a devout Muslim, but he belonged to a sect that most Muslims consider heretic, and for years was not allowed to return to Pakistan.)

Dawkins treats Islam as just another deplorable religion, but there is a difference. The difference lies in the extent to which religious certitude lingers on in the Muslim world, and in the harm it does. Dawkins's even-handedness is well-intentioned, but it is misplaced. I share Dawkins's lack of respect for all religions, but in our times it is folly to disrespect them all equally.

22

To the Postdocs

The postdoctoral research associateship is a wonderful invention of the American academic community, but the experience can be frightening to those who hold these positions. These "postdocs" are newly minted Ph.D.s, given an opportunity to do research at a university or national laboratory for a term of two or three years, free of teaching duties, but without any expectation that they will be considered for promotion there to a more permanent faculty or staff position. The unspoken message that comes with appointment as a postdoc is "O.K., you did well enough as a graduate student to get this job, and now you have a chance to show what kind of research you can do on your own initiative. From now on, no one will care where you went to school or how well you did there, but only what kind of research you do now. You have two or three years, and then you will have to try to get another job, as an assistant professor or equivalent that can lead to something more permanent. Don't just sit there; discover something."

It is natural that postdocs feel a certain anxiety. In the the autumn of 2006 the Department of Physics at the University of Texas invited a number of postdocs from other American universities to Austin to a workshop to discuss their current work with each other and with our faculty. I was asked to give an after-dinner talk to the postdocs, and decided that a little encouragment might not come amiss. The talk seemed to be well received, and I submitted the following written version to *Physics Today*, which published it in March 2007.

In thinking over what I would say this evening to a group of talented young postdocs at the beginning of their research careers, I naturally thought back to how things seemed to my generation of physicists when we were first starting out in research. Many of us were worried about how difficult it seemed to make progress in the state that physics was in then. (I am remembering how things were in my own area of physics, the theory of particles and fields, but I would not be surprised if similar remarks applied in other areas.) We had a theory of weak nuclear forces[1] that gave nonsense when pushed beyond the lowest order of approximation. The strong nuclear forces[2] were even more puzzling. We had no reason to believe in any particular theory of these forces, and no way to calculate the consequences of a theory of strong nuclear forces even if we did believe in it. Some people thought that the path to understanding the strong forces led through the study of the analytic structure of scattering amplitudes as functions of several kinematic variables.[3] That approach really depressed me because I knew that I could never understand the theory of more than one complex variable. So I was pretty worried about how I could do research working in this mess.

1. Added note: These are the forces that, among other things, are responsible for radioactive decay processes in which a neutron or proton inside an atomic nucleus changes to a proton or neutron, emitting other energetic particles like electrons and neutrinos.

2. Added note: These are the forces that hold quarks together inside neutrons and protons and other particles, and that hold neutrons and protons together inside atomic nuclei.

3. Added note: This is hard to explain in nonmathematical terms. Scattering amplitudes are numerical quantities whose squares give the probabilities of the various things that can happen when a beam of particles is scattered by other particles. They depend on what are called kinematic variables, such as the energies and directions of motion of the particles in the initial and final states of the scattering process. The "analytic structure" is a specification of how smooth is this dependence, when the range over which the kinematic variables are allowed to vary is extended to include complex numbers—that is, numbers involving the square roots of negative numbers.

I have to confess that on top of my pessimism, I felt a sense of envy of the previous generation of theorists. Perhaps many of my generation shared this feeling. We saw that the generation of Freeman Dyson, Richard Feynman, Julian Schwinger, and Sin-Itiro Tomonaga in the late 1940s had at hand the twenty-year-old theory of quantum electrodynamics.[4] It seemed to me that all they had needed to do was to recognize how the measured values of the electron's mass and charge are related to the symbols m and e appearing in the field equations. Once this was sorted out, Schwinger could easily—or so it seemed to me—have calculated the magnetic moment[5] of the electron to four decimal places. It all seemed much easier than the puzzles faced by our generation of physicists.

Of course, we were wrong to envy the previous generation, and for two reasons.

For one thing, I dare say that every generation of physicists has envied its predecessors. I know that some theorists of the generation of Feynman and Schwinger regretted that they had not worked in the 1920s, when quantum mechanics was discovered. After all, what was so hard about guessing the Schrödinger equation[6] and then solving it for the spectrum of the hydrogen atom? I suppose that in the 1920s Werner Heisenberg, Wolfgang Pauli, Erwin Schrödinger, and Paul Dirac must have envied Albert Einstein, who in developing General Relativity in 1915 only had to deal with field equations, without worrying about the complications of quantum mechanics. And who in the world could Einstein have envied? Clearly, no one but Isaac Newton! Sure enough, Einstein's

4. Added note: Quantum electrodynamics is the quantum mechanical theory of electrons, positrons (antielectrons), and photons (particles of light).
5. Added note: The magnetic moment is a number that gives the strength of the magnetic field produced by the electron's spin.
6. Added note: The Schrödinger equation for an atomic state of definite energy governs the spatial variation of the wave function of the particles in the state, the quantity whose square gives the probability of finding the particles in one or another position. This equation has properly behaved solutions only when the total energy of the atom takes certain discrete values, known as the spectrum of the atom.

foreword to the 1931 edition of Newton's *Opticks* begins, "Fortunate Newton, happy childhood of science!"

Also, in every generation, those who thought that the problems of their predecessors were easier than their own had been wrong. It took courage for Dyson, Feynman, Schwinger, and Tomonaga to take quantum electrodynamics seriously. In the late 1940s, it was generally thought that quantum electrodynamics was only a low-energy approximation, which could not be trusted at energies above about a few million electron volts, and which had to be replaced by something entirely new. When Schrödinger wrote down his equation in 1925, he had no idea what the wave function meant; that had to wait until Max Born's work on scattering theory a few years later. And Newton didn't just invent a specific law of motion and a specific law of gravitational force—he had to invent the whole idea of dynamical equations. Before Newton, there had only existed a limited mathematical kinematics,[7] worked out in the Middle Ages by Jean Buridan and others, and the worthless philosophical dynamics of Aristotle.

So the moral of my tale is not to despair at the formidable difficulties that you face in getting started in today's research. In fact, the opportunities for progress lie in just those areas of physics that seem most messy. My generation is not handing over to yours a clear set of tasks, like the problems in a physics textbook, but when has it ever been clear what is the next thing to be done? You are far better trained mathematically than any previous generation of physicists, and you have at your disposal tools, from personal computers to artificial satellites, beyond the dreams of earlier scientists. You'll have a hard time, as we did, but you'll do OK.

7. Added note: This isn't quite true. A little before Newton, Christiaan Huygens had found an important formula for the centripetal acceleration of a body travelling in a circle at constant speed.

23

Science or Spacemen

No field of science owes as much to amateurs as astronomy. With binoculars and modest telescopes they patrol the sky for comets and variable stars that would be missed by the professional astronomers' large telescopes, with their narrow fields of view. Even more important, the amateur astronomers form a political constituency that is interested in the results of astronomical research, and generally supports government spending on this work.

In the United States the community of amateur astronomers (plus not a few professionals) is served by the magazine *Sky and Telescope*. So it was natural that when I wanted to muster support for a project to study cosmic rays from the International Space Station, I would submit an opinion piece to this magazine.

The key instrument involved in this project is the Alpha Magnetic Spectrometer (AMS), which uses the techniques of high-energy physics to diagnose the nature of the particles making up cosmic rays. The AMS is ready to use, and only needs to be brought up to the International Space Station. My support for this project was enlisted in 2007 by its leader, my old friend the physicist Sam Ting. I learned from him that NASA, after promising to use the Space Shuttle to bring the AMS up to the International Space Station, was reneging on this promise. After President Bush announced his plans to send astronauts back to the Moon and then on to Mars, NASA had determined to use the Shuttle only to complete the Space Station, and to use the Space Station only to support the technology of manned space flight. It seemed to me that this was another harmful consequence of the tendency of NASA to devote resources to grandiose missions of manned space flight, like the Moon-Mars mission. I had already expressed my negative views of manned space flight a few years earlier—the article reprinted in chapter 14 of this collection—and now I wanted to use the case of the AMS as an example of the harm done by the emphasis on manned space flight. My article was published by *Sky and Telescope* in May 2008.

———

For a half century, NASA has emphasized human space flight above all else. Undoubtedly, some people are excited just seeing astronauts in space, but this is an awfully expensive spectator sport.

Human space exploration is often justified for its contribution to science and technology. Yes, putting people in space is essential to studying of the effects of microgravity on humans, but that's not an interesting issue unless there is some other reason to put people into space. NASA missions have fostered great advances in astronomy. But all those discoveries have come from unmanned robotic craft such as the Hubble Space Telescope, the Cosmic Background Explorer, and the Wilkinson Microwave Anisotropy Probe.

Frankly, there's not much people can do in space that instrumented satellites and robots cannot do better and at much lower cost. People radiate heat and bump into things. They need air and food and water, and they insist on coming back to Earth.

Most astronomers have been unwilling to criticize NASA's continued emphasis on human space flight. Their rationale has been that having people in space appeals to the public, which increases the space agency's funding, some of which supports real science. But this tacit bargain is now unravelling. President Bush's enormously costly plan for sending humans to the Moon and Mars is forcing NASA to cut way back on various science missions, from monitoring climate change to studying the cosmic microwave background.

There is one particularly egregious example of NASA's willingness to sacrifice science in the cause of rocketing people into space. An international collaboration headed by the Nobel laureate Samuel

Ting of MIT has built an impressive instrument known as the Alpha Magnetic Spectrometer (AMS) for studying high-energy primary cosmic rays from space. The AMS is destined for the International Space Station, not because astronauts are needed to operate it, but because the Space Station is a stable platform above Earth's atmosphere with lots of available power. As long as the Space Station is up there, we might as well use it to do some significant science.

The AMS is just about ready to fly. Roughly $1.5 billion has been spent to construct it, mostly by a European consortium, and NASA agreed to deliver it to the Space Station. But in the aftermath of *Columbia*'s tragic loss in 2003, NASA managers decided that work on the Space Station would be limited to supporting the President's vision for human space exploration. Consequently, the agency is now unwilling to devote a quarter of the payload of any one of the remaining Space Shuttle flights to put the AMS on the Space Station. I gather NASA wants to use the AMS as motivation for an additional Shuttle flight, but this is being blocked by Congress. And although a senior official of NASA has told me that this mission could be included in the exploration budget, I fear in the end astronomy would bear the cost.

It is time for astronomers and lovers of astronomy to push back against the repeated decisions of NASA and the Bush administration to pursue human space flight at the expense of real science. Some will counter that keeping astronauts aloft is what the public wants. This might have once been true, but I wonder if it still is. Today's Web-savvy space buffs may be more receptive to results from remote sensors than were their predecessors. There's been great public interest in results from the Hubble telescope and from the rovers Spirit and Opportunity on Mars.

No doubt it will be tremendously exciting once astronauts return to the Moon and go on to Mars—at first. But, as experience has shown, the public will soon become disenchanted when it becomes clear that there's really not much for astronauts to do once they get there.

Since this article was published, I have learned that NASA may add additional Shuttle flights to its program, and use one of them to bring the AMS up to the Space Station. I am not sure whether this is a good thing. I first heard of this possibility when I gave the opening speech by video conferencing to the Alpha Magnetic Spectrometer Technical Interchange Meeting, at the Center for Advanced Space Studies of the Lunar and Planetary Institute in Houston in October 2007. In the discussion period after my talk, I got into an argument about adding extra Shuttle flights with Congressman Nicholas Lampson, who then represented the district in which the Johnson Manned Space Flight Center is located. It seemed to me that there was plenty of room on scheduled Shuttle flights to take the AMS up to the Space Station. Adding additional Shuttle flights would be good for the Manned Space Flight Center, but I still fear that it will be likely to draw more funds away from scientific research.

24

Israel and the Liberals

The year 2008 marked the sixtieth anniversary of the founding of the state of Israel. The magazine *Tikkun* commemorated this anniversary with a special section, "Israel at 60," in their May/June 2008 issue. To my surprise, I was one of those invited to contribute to this section. I was surprised because I had taken strong pro-Israel stands in earlier articles in the *New Republic* (reprinted in my earlier essay collection, *Facing Up*) and in the *Times Higher Education Supplement* (chapter 20 in the present collection). It seemed to me that, though I would not call *Tikkun* anti-Israel, it frequently seeks solutions to the conflict between Israel and the Arabs in things that Israel rather than the Arabs ought to do differently. I anticipated that I might have trouble getting my contribution published without excessive editorial changes. I needn't have worried—the editors at *Tikkun* left my article pretty much as I wrote it.

The greatest miracle of our time is the rebirth of Israel in its ancient home. And, with it, the transformation of an inhospitable landscape into the lovely land of Israel, with its tree-lined streets, cafes, universities, its optimistic people, its liberated women, its liberal democracy and rule of law. The continuing miracle is the survival of this remnant of a life-loving people on a sliver of land, despite repeated attacks from hostile Arab armies. Edward Gibbon wrote of the condition of European Jews in the early Middle Ages that "they might be oppressed without danger, as they had lost the use, and even the remembrance of arms." Miraculously, in the twentieth century Jews learned again to use arms to defend themselves.

But though Israel has survived, it faces continued hostility, and not only from Arab irredentists, from politicians and businessmen who curry favor with oil-rich Muslim countries, and from ordinary old-fashioned anti-Semites. The most horrifying development of our time, with respect to Israel, is the worldwide conversion of many of my fellow liberal intellectuals—academics, artists, writers, labor leaders, "enlightened" clergy—to irrational hatred of Israel, the only liberal democracy in the Middle East and the only country in that region where any of us could bear to live. They sit in world forums, weirdly denouncing Israel as the worst violator of human rights, the greatest threat to world peace. They demand that Israel cease its checkpoints and walls and its other attempts to save Jewish lives. The most shameful participants in all this are Jewish and Israeli "liberals," rushing to show the world that they are not the bad Jews, but the good ones, more anti-Israel than thou.

The ostensible root cause of this hostility is the supposed "occupation" of Palestinian land. Often this takes the form of a demand that Israel return to its 1967 borders. Israel has already made a complete withdrawal from Gaza and handed over most of the West Bank to the Palestinian Liberation Organization (PLO). Can anyone really suppose that clearing Jews from the small remaining area they inhabit and ending Israeli military patrols in the West Bank would bring peace? Hamas, Hezbollah, and their patrons in Muslim countries like Iran have made it clear that they will not be satisfied with anything but the annihilation of Israel. Supposed "moderates" in the PLO and some Muslim countries now say they accept the existence of Israel, but what line would they take if Israel returned to its strategically vulnerable 1967 borders? Remember, the PLO was founded in 1964, when no part of Gaza or the West Bank was in Israeli hands, and when "liberation" could only mean the annihilation of Israel.

Not only is the demand for total Israeli withdrawal from the West Bank unrealistic—it is unjust. Countries that are attacked, and defeat their attackers, are not usually blamed if they then take territory from their attacker that will help to protect them from further attacks. This is what the Soviet Union did after World War II when it annexed East Prussia, and not even most Germans now condemn them for it. (Of course, Russia did not have the problem of ruling Germans in the land it seized, because it deported them all.) By these standards, Israel deserves to hang on to whatever part of the West Bank it needs for its security.

Then there are "liberals" who deny the legitimacy of Israel itself, and ask why there should not be a multiethnic state in the whole of Palestine, subject to a majority that happens to be Arab? They ignore the historical evidence that Jews cannot live freely in a country dominated by Arabs. Jews living among Arabs were subject to repeated Arab massacres, as in Hebron in 1929, and where Jews in Arab countries were tolerated, they were never more than second-class citizens, a status dictated in the Qu'ran. Jews are now prohibited from living in Saudi Arabia, and have had to flee other Arab countries. Although world law established all of Palestine as the Jewish national home, after World War I the British handed most of it (the part today known as Jordan)[1] to England's Hashemite allies, who had no connection to Jordan. But the Jews have real connections to Palestine, their ancient home, where thousands of Jews have always lived since the Roman era. Even after the handover to Jordan, the United Nations voted to partition the remnant, leaving only half of it to the Jews. Yet the Jews accepted their half-sliver of land, and created a beautiful country.

It is time for liberals and intellectuals, Jewish and gentile alike, to return to our true liberal values. We should acknowledge that hatred of the Jewish state is the functional equivalent of hatred of Jews. Before it is too late, we must do what we can to secure the continued survival of Jews in their own country.

1. Added note: The 1911 edition of the *Encyclopedia Brittanica*, which presumably reflects at least the British view at about this time, describes the river Jordan as marking "a line of delimitation between Western and Eastern Palestine," not as the border between Palestine and somewhere else.

25

Without God

In June 2007 I received an e-mail message from Howard Georgi, a good friend at the Harvard Physics Department with whom I had collaborated when I was at Harvard. Howard had become president of the Harvard chapter of Phi Beta Kappa and was now inviting me to be the orator at the Phi Beta Kappa Literary Exercises during the Commencement Week in June 2008. I knew that these Phi Beta Kappa orations were a grand affair. They have been given since the eighteenth century by New England luminaries including John Quincy Adams, Edward Everett, Henry Wadsworth Longfellow, Josiah Royce, and Oliver Wendell Holmes, then in the twentieth century by other Americans including Woodrow Wilson and Walter Lippmann, and more recently even by a few scientists such as Edwin Land and Lewis Thomas. Not being immune to the attraction of participating in such a tradition, I immediately accepted.

For this oration, I decided to rework a lecture on the tension between science and religion that I had given as part of a series of Messenger Lectures at Cornell University in May 2007. I had a chance to polish it further when I debated the molecular biologist Francis Collins in the Trotter Prize exercises at Texas A&M University in March 2008. For the Phi Beta Kappa oration I thought it was appropriate to pay my respects to Ralph Waldo Emerson, who in 1837 had given what became the most famous of all the Phi Beta Kappa orations, published as *The American Scholar.*

The Phi Beta Kappa Literary Exercises in June 2008 did indeed turn out to be a grand occasion. They were held in Harvard's Sanders Theater, and presided over by the new President of Harvard, Drew Gilpin Faust. The exercises included a reading by the year's Phi Beta Kappa poet, Carl Phillips, and a concert of hymns by the all-male Harvard Glee Club. When in my oration I mentioned that having no religious belief need not prevent us from enjoying religious poetry, I added that even I can enjoy the beautiful hymns we had heard, and that nothing could have improved their performance, unless the Harvard Glee Club were finally to include women's voices along with the men's. That was the one point in my oration when I was interrupted by applause, along with some boos.

The following article is an edited version of my talk at Harvard, leaving out the dig at the Harvard Glee Club. It was published in the *New York Review of Books* in September 2008. The article evoked unusually many responses from readers, some sent to the *New York Review,* and many others directly to me. In what follows, I have incorporated some corrections pointed out by readers, and expanded the article in places where readers' comments showed that additions or clarifications were needed.

In his celebrated 1837 Phi Beta Kappa oration at Harvard, titled *The American Scholar,* Ralph Waldo Emerson predicted that a day would come when America would end what he called "our long apprenticeship to the learning of other lands." His prediction came true in the twentieth century, and in no area of learning more so than in science. This surely would have pleased Emerson. When he listed his heroes he would generally include Copernicus and Galileo and Newton along with Socrates and Jesus and Swedenborg. But I think that Emerson would have had mixed feelings about one consequence of the advance of science here and abroad—that it has led to a widespread weakening of religious belief. Emerson was hardly orthodox—according to Herman Melville, Emerson felt "that had he lived in those days when the world was made, he might have offered some valuable suggestions"—but he was for a while a Unitarian minister, and he usually found it possible to speak favorably of the Almighty. Emerson grieved over what he saw in his own time as a weakening of belief, as opposed to mere piety and church-going, in America and even more so in England, though I can't say that he attributed it to the advance of science.

The idea of a conflict between science and religion has a long pedigree. According to Edward Gibbon, it was the view of the Byzantine church that "the study of nature was the surest symptom of an unbelieving mind." Perhaps the best known portrayal of this conflict is a book published in 1896 by Cornell's first president, Andrew Dickson White, with the title "A History of the Warfare of Science with Theology in Christendom."

In recent times there has been a reaction against talk of warfare between science and religion. White's "conflict thesis" was attacked in a 1986 paper by Bruce Lindberg and Ronald Numbers, both well-known historians of science, who pointed out many flaws in White's scholarship. The Templeton Foundation offers a large prize to those who argue that there is no conflict between science and religion. Some scientists take this line because they want to protect science education from religious fundamentalists. The late Stephen Gould argued that there could be no conflict between science and religion, because science deals only with facts and religion only with values. This certainly was not the view held in the past by most adherents of religion, and it is a sign of the decay of belief in the supernatural that many today who call themselves religious would agree with Gould.

Let's grant that science and religion are not incompatible—there are after all some (though not many) excellent scientists like Charles Townes and Francis Collins who have strong religious beliefs. Still, I think that between science and religion there is, if not an incompatibility, at least what the philosopher Susan Haack has called a tension, that has been gradually weakening serious religious belief, especially in the West, where science is most advanced. Here I would like to trace out some of the sources of this tension, and then offer a few remarks about the very difficult question raised by the consequent decline of belief, the question of how it will be possible to live without God.

1

I do not think that the tension between science and religion is primarily a result of contradictions between scientific discoveries and specific religious doctrines. This is what chiefly concerned White, but I think he was looking in the wrong direction. Galileo remarked in his famous letter to Grand Duchess Christina that "the intention of the Holy Ghost is to teach us how to go to heaven, not how heaven goes," and this was not just his opinion; he was quoting a prince of the Church, Cardinal Baronius, the Vatican librarian.

Contradictions between scripture and scientific knowledge have arisen again and again, and have generally been accommodated by the more enlightened among the religious. For instance, there are verses in both the Old and New Testament that seem to show that the Earth is flat, and as noted by Copernicus (quoted in the same Letter to Christina) these verses led some early Church Fathers like Lactantius to reject the Greek understanding that the Earth is a sphere, but educated Christians long before the voyages of Columbus and Magellan had become comfortable with the spherical shape of the Earth. Dante found the interior of the spherical Earth a convenient place to store sinners.

What was briefly a serious issue in the early Church has today become a parody. The astrophysicist Adrian Melott of the University of Kansas, in a fight with zealots who wanted equal time for creationism in the Kansas public schools, founded an organization called FLAT (Families for Learning Accurate Theories). His society parodied creationists by demanding equal time for flat Earth geography, arguing that children should be exposed to both sides of the controversy over the shape of the Earth.

But if the direct conflict between scientific knowledge and specific religious beliefs has not been so important in itself, there are at least four sources of tension between science and religion that *have* been important.

The first source of tension arises from the fact that religion originally gained much of its strength from the observation of mysterious phenomena—thunder, earthquakes, disease—that seemed to require the intervention of some divine being. There was a nymph in every brook, and a dryad in every tree. But as time passed more and more of these mysteries have been explained in purely natural ways. Explaining this or that about the natural world does not of course rule out religious belief. But if people believe in God because no other explanation seems possible for a whole host of mysteries, and then over the years these mysteries are one by one resolved naturalistically, then a certain weakening of belief can be expected. It is no accident that the advent of widespread atheism and agnosticism and deism among the educated in the eighteenth century followed hard upon the birth of modern science in the previous century.

From the beginning, the explanatory power of science worried those who valued religion. Plato was so horrified at the attempt of Democritus and Leucippus to explain nature in terms of atoms without reference to the gods (even though they did not get very far with this) that in Book Ten of the *Laws* he urged five years of solitary confinement for those who deny that the gods exist or that they care about humans, with death to follow if the prisoner is not reformed. Isaac Newton, offended by the naturalism of Descartes, also rejected the idea that the world could be explained without God. He argued for instance in a letter to Richard Bentley that no explanation but God could be given for the distinction we observe between bright matter, the Sun and stars, and dark matter, like the Earth. This is ironic, because of course it was Newton and not Descartes who was right about the laws of motion. No one did more than Newton to make it possible to work out thoroughly nontheistic explanations of what we see in the sky, but Newton himself was not in this sense a Newtonian.

Of course, not everything has been explained, nor will ever be. The important thing is that we have not observed anything that seems to require supernatural intervention for its explanation. There are some today who cling to the remaining gaps in our understanding (such as our ignorance about the origin of life) as evidence for God. But as time passes and more and more of these gaps are filled in, their position gives an impression of people desperately holding on to outmoded opinions.

The problem for religious belief is not just that science has explained a lot of odds and ends about the world. There is a second source of tension, that these explanations have cast increasing doubt on the special role of man, as an actor created by God to play a starring role in a cosmic drama that He has authored. We have had to accept that our home, the Earth, is just another planet circling the Sun; our Sun is just one of a hundred billion stars in a galaxy that is just one of billions of visible galaxies; and it may be that the whole expanding cloud of galaxies is just a small part of a much larger multiverse, most of whose parts are utterly inhospitable to life. As Richard Feynman has said, "the theory that it's all arranged as a stage for God to watch man's struggle for good and evil seems inadequate."

Most important so far has been the discovery by Darwin and Wallace that humans arose from earlier animals through natural selection acting on random heritable variations, with no need for a divine plan to explain the advent of humanity. This discovery led some, including Darwin, to lose their faith. It's not surprising that of all the discoveries of science, this is the one that continues most to disturb religious conservatives. I can imagine how disturbed they will feel in the future, when at last scientists learn how to understand human behavior in terms of the chemistry and physics of the brain, and nothing is left that needs to be explained by an immaterial soul.

Note that I refer here to *behavior,* not consciousness. Something purely subjective, like how we feel when we see the color red or discover a physical theory, seems so different from the objective world described by science that it is difficult to see how they can ever come together. As Colin McGinn has said, "The problem is how to integrate the conscious mind with the physical brain—how to reveal a unity beneath this apparent diversity. That problem is very hard, and I do not believe anyone has any good ideas about how to solve it." On the other hand, both brain activity and be-havior (including what we say about our feelings) are in the same world of objective phenomena, and I know of no intrinsic obstacle to their being integrated in a scientific theory, though it is clearly not going to be easy. This does not mean that we can or should forget about consciousness, and like B. F. Skinner with his pigeons concern ourselves only with behavior. We know, as well as we know anything, that our behavior is partly governed by our con-sciousness, so understanding behavior will necessarily require work-ing out a detailed correspondence between objective behavior and subjective consciousness. This may not tell us how one arises from the other, but at least it will confirm that there is nothing super-natural about the mind.

Some nonscientists seize on certain developments in modern physics that suggest the unpredictability of natural phenomena, such as the advent of quantum mechanics or chaos theory, as sig-nals of a turn away from determinism, of the sort that would make an opening for divine intervention or an incorporeal soul. These theories have forced us to refine our view of determinism, but not I think in any way that has implications for human life.

A third source of tension between science and religious belief has been more important in Islam than in Christianity. Around 1100 the Sufi philosopher Abu Hamid al-Ghazali argued against the very idea of laws of nature, on the grounds that any such law would put God's hands in chains. According to al-Ghazali, a piece of cotton placed in a flame does not darken and smolder because of the heat of the flame, but because God wants it to darken and smolder. Laws of nature could have been reconciled with Islam, as a summary of what God usually wants to happen, but al-Ghazali did not take that path.

Of course, al-Ghazali did not think that there was no relation at all between cause and effect. I can't imagine that he went around putting his hand in fires. But al-Ghazali did not think that explor-ing these relations was a good thing. He compared astronomy and mathematics to wine. Wine strengthens the body, but is nevertheless forbidden; similarly, astronomy and mathematics strengthen the mind, but "we nevertheless fear that one might be attracted through them to doctrines that are dangerous."

Al-Ghazali is often described as the most influential Islamic philosopher. I wish I knew enough to judge how great was the effect on Islam of his rejection of science. Certainly many Muslim scholars have emphasized his impact, from Ibn Rushd in the twelfth century to Syed Ameer Ali and Pervez Hoodbhoy in the twentieth. At any rate, science in Muslim countries, which had led the world in the ninth and tenth centuries, went into a decline in the century or two after al-Ghazali. As a portent of this decline, in 1194 the Ulama of Cordoba burned all scientific and medical texts.

Nor has science revived in the Islamic world. There are talented scientists who have come to the West from Islamic countries and here do work of great value, among them the Pakistani Muslim physicist Abdus Salam, who in 1979 became the first Muslim scientist to be awarded the Nobel Prize, for work he did in England and Italy. But in the past forty years I have not seen any paper in the areas of physics or astronomy that I follow that was written in an Islamic country and was worth reading. Thousands of scientific papers are turned out in these countries, and perhaps I missed something. Still, in 2002 the periodical *Nature* carried out a survey of science in Islamic countries, and found just three areas in which the Islamic world produced excellent science, all three directed toward applications rather than basic science. They were desalination, falconry, and camel breeding.

Something like al-Ghazali's concern for God's freedom surfaced for a while in Christian Europe, but with very different results. In Paris and Canterbury in the thirteenth century there was a wave of condemnations of those teachings of Aristotle that seemed to limit the freedom of God to do things like create a vacuum or make several worlds or move the heavens in straight lines. The influence of Thomas Aquinas and Albertus Magnus saved the philosophy of Aristotle for Europe, and with it the idea of laws of nature. But although Aristotle was no longer condemned, his authority had been questioned, which was fortunate, since nothing could be built on his physics. Perhaps it was the weakening of Aristotle's authority by reactionary churchmen that opened the door to the first small steps toward finding the true laws of nature at Paris and Oxford in the fourteenth century.

There is a fourth source of tension between science and religion that may be the most important of all. Traditional religions generally rely on authority, whether the authority is an infallible leader, such as a prophet or a pope or an imam, or a body of sacred writings, a Bible or a Koran. Perhaps Galileo did not get into trouble solely because he was expressing views contrary to scripture, but because he was doing so independently, rather than as a theologian acting within the Church.

Of course, scientists rely on authorities, but of a very different sort. If I want to understand some fine point about the General Theory of Relativity, I might look up a recent paper by an expert in the field. But I would know that the expert might be wrong. One thing I probably would *not* do is to look up the original papers of Einstein, because today any good graduate student understands General Relativity better than Einstein did. We progress. Indeed, in the form in which Einstein described his theory it is today generally regarded as only what is known in the trade as an effective field theory; that is, it is an approximation, valid for the large scales of distance for which it has been tested, but not under very cramped conditions, as in the early big bang.

We have our heroes in science, like Einstein, who was certainly the greatest physicist of the past century, but for us they are not infallible prophets. For those who in everyday life respect independence of mind and openness to contradiction, traits that Emerson admired—especially when it came to religion—the example of science casts an unfavorable light on the deference to authority of traditional religion. The world can always use heroes, but could do with fewer prophets.

The weakening of religious belief is obvious in western Europe, but it may seem odd to talk about this happening in America. No one who expressed doubt about the existence of God could possibly be elected President of the United States. Nevertheless, though I don't have any scientific evidence on this point, on the basis of personal observation it seems to me that, while many Americans fervently believe that religion is a good thing, and get quite angry when it is criticized, even those who feel this way often do not have much in the way of clear religious *belief.* Occasionally I have found myself talking with friends who identify themselves with some organized religion about what they think of life after death, or of the nature of God, or of sin. Most often I've been told that they do not know, and that the important thing is not what you believe, but how you live. I've even heard this from a Catholic priest. I welcome the sentiment, but it's quite a retreat from religious belief.

Though I can't prove it, I suspect that when Americans are asked in polls whether they believe in God or angels or heaven or hell they feel that it is a pious duty to say that they do, whatever they actually believe. In *Sacred and Secular: Religion and Politics Worldwide,* Pippa Norris and Ronald Inglehart report that by actually counting the participants at church services, it is found that Americans have a tendency to exaggerate their church attendance, "due to a social desirability bias concerning churchgoing in American culture." If such problems arise in using surveys to judge something tangible like church attendance, how much more are they likely to intrude on surveys of belief? And of course hardly anyone today in the West seem to have even the slightest interest in the great controversies—Arians versus Athanasians, monophysites versus monothelites, justification by faith or by works—that used to be taken so seriously that they set Christians at each other's throats.

I have been emphasizing religious belief here, the belief in facts about God or the afterlife, though I am well aware that this is only one aspect of the religious life, and for many not the most important part. Perhaps I emphasize belief, because as a physicist I am professionally concerned with finding out what is true, not what makes us happy or good. For many people, the important thing about their religion is not a set of beliefs but a host of other things: a set of moral principles; rules about sexual behavior, diet, observance of holy days, and so on; rituals of marriage and mourning; and the comfort of affiliation with fellow believers, which in extreme cases allows the pleasure of killing those who have different religious affiliations.

For some there is also a sort of spirituality that Emerson wrote about, and which I don't understand, often described as a sense of union with nature or with all humanity, that doesn't involve any specific beliefs about the supernatural. Spirituality is central to Buddhism, which does not call for belief in God. Even so, Buddhism has historically relied on belief in the supernatural, specifically in reincarnation. It is the desire to escape the wheel of rebirth that drives the search for enlightenment. The heroes of Buddhism are the Bodhisativas, who having attained enlightenment, nevertheless return to life in order to show the way to a world shrouded in darkness. Perhaps in Buddhism too there has been a decline of belief. A recent book by the Dalai Lama barely mentions reincarnation, and Buddhism is now in decline in Japan, the Asian nation that has best matched the West in scientific research.

The various uses of religion may keep it going for a few centuries even after the disappearance of belief in anything supernatural, but I wonder how long religion can last without a core of belief in the supernatural, when it isn't about anything external to human beings. To compare great things with small, people may go to college football games mostly because they enjoy the cheerleading and marching bands, but I doubt if they would keep going to the stadium on Saturday afternoons if the only things happening there were cheerleading and marching bands, without any actual football, so that the cheerleading and the band music were no longer *about* anything.

2

It is not my purpose here to argue that the decline of religious belief is a good thing (although I think it is), or to try to talk anyone out of their religion, as in eloquent recent books by Richard Dawkins, Sam Harris, and Christopher Hitchens. So far in my life, in arguing for spending more money on scientific research and higher education, or against spending on ballistic missile defense or sending people to Mars, I think I have kept a perfect record of never having changed anyone's mind. Rather, I want just to offer a few opinions, on the basis of no expertise whatever, for those who have already lost their religious beliefs, or who may be losing them, or fear that they will lose their beliefs, about how it is possible to live without God.

First, a warning: we had better beware of substitutes. It has often been noted that the greatest horrors of the twentieth century were perpetrated by regimes—Hitler's Germany, Stalin's Russia, Mao's China—that while rejecting some or all of the teachings of religion, copied characteristics of religion at its worst: infallible leaders, sacred writings, mass rituals, the execution of apostates, and a sense of community that justified exterminating those outside the community.

When I was an undergraduate I knew a theologian, Will Herberg, who worried about my lack of religious faith. He warned me that we must worship God, because otherwise we would start worshipping each other. He was right about the danger, but I would suggest a different cure: we should get out of the habit of worshipping anything.

Don't worry, I'm not going to say that it's easy to live without God, that science is all you need. For a physicist, it is indeed a great joy to learn how we can use beautiful mathematics to understand the real world. We struggle to understand nature, building a great chain of research institutes, from the Museum of Alexandria and the House of Wisdom of Baghdad to today's CERN and Fermilab. But we know that we will never get to the bottom of things, because whatever theory unifies all observed particles and forces, we will never know why it is that theory that describes the real world and not some other theory.

Worse, the worldview of science is rather chilling. Not only do we not find any point to life laid out for us in nature, no objective basis for our moral principles, no correspondence between what we think is the moral law and the laws of nature, of the sort imagined by philosophers from Anaximander and Plato to Emerson. We even learn that the emotions that we most treasure, our love for our wives and husbands and children, are made possible by chemical processes in our brains that are what they are as a result of natural selection acting on chance mutations over millions of years. And yet we must not sink into nihilism or stifle our emotions. At our best we live on a knife-edge, between wishful thinking on one hand and, on the other, despair.

What, then, can we do? One thing that helps is humor, a quality not abundant in Emerson. Just as we laugh with sympathy but not scorn when we see a one-year-old struggling to stay erect when she takes her first steps, we can feel a sympathetic merriment at ourselves, trying to live balanced on a knife-edge. In some of Shakespeare's greatest tragedies, just when the action is about to reach an unbearable climax, the tragic heroes are confronted with some "rude mechanical" offering comic observations: a gravedigger, or a doorkeeper, or a pair of gardeners, or a man with a basket of figs. The tragedy is not lessened, but the humor puts it in perspective.

Then there are the ordinary pleasures of life, which have been despised by religious zealots, from Christian anchorites in the Egyptian deserts to today's Taliban and Mahdi Army. Visiting New England in early June, when the rhododendrons and azaleas are blazing away, reminds one how beautiful spring can be. And let's not dismiss the pleasures of the flesh. We who are not zealots can rejoice that when bread and wine are no longer sacraments, they will still be bread and wine.

There are also the pleasures brought to us by the high arts. Here I think we *are* going to lose something with the decline of religious belief. Much great art has arisen in the past from religious inspiration. For instance, I can't imagine the poetry of George Herbert or Henry Vaughan or Gerard Manley Hopkins being written without sincere religious belief. But nothing prevents those of us who have no religious belief from enjoying religious poetry (or cathedral architecture or requiem masses), any more than not being English prevents Americans from enjoying the patriotic speeches in *Richard II* or *Henry V.*

We may be sad that no more great religious poetry will be written in the future. We see already that not much significant English-language poetry written in the past few decades owes anything to belief in God, and some cases where religion does enter, in the case of poets like Stevie Smith or Philip Larkin, it is the rejection of religion that provides their inspiration. But of course very great poetry can be written without religion. Shakespeare provides an example; none of his work seems to me to show the slightest hint of serious religious inspiration.

Shakespeare of course wrote in a religious time, and about religious times, so churchmen necessarily appear as characters in Shakespeare's plays, and his language draws on many words and ideas from a Christian tradition. But his clergy are generally worldly schemers, like Winchester in *Henry VI* or Canterbury and Ely in *Henry V* or Wolsey in *Henry VIII*, or they are marplots, like Friar Lawrence, or prigs, like the priest who refuses to bury Ophelia in hallowed ground. More significant, few if any of his characters are motivated by religious belief. It's true that Hamlet tells himself that he would commit suicide if the Almighty had not fixed His canon against self-slaughter, and that he would kill Claudius if the king were not at his prayers, but aren't these just more of his excuses for delay? Where Hamlet seems most sincere, as in his meditation on the skull of Yorick, he is least Christian. When Shakespeare brings in supernatural characters, they are pagan: fairies and sorcerers. Given Ariel and Prospero, we see that poets can do without angels and prophets.

I do not think we have to worry that giving up religion will lead to a moral decline. There are plenty of people without religious faith who live exemplary moral lives (as for example, me), and though religion has sometimes inspired admirable ethical standards, it has also often fostered the most hideous crimes. Anyway, belief in an omnipotent omniscient creator of the world does not in itself have any moral implications—it's still up to you to decide whether it is right to obey His commands. For instance, even someone who believes in God can feel that Abraham in the Old Testament was wrong to obey God in agreeing to sacrifice Isaac, and that Adam in *Paradise Lost* was right to disobey God and follow Eve in eating the apple, so that he could stay with her when she was driven from Eden. The young men who flew airplanes into buildings in the United States or exploded bombs in crowds in London or Madrid or Tel Aviv were not just stupid in imagining that these were God's commands; even thinking that these were His commands, they were evil in obeying them.

The more we reflect on the pleasures of life, the more we miss the greatest consolation that used to be provided by religious belief: the promise that our lives will continue after death, and that in the afterlife we will meet the people we have loved. As religious belief weakens, more and more of us know that after death there is nothing.

Cicero offered comfort in *De Senectute* by arguing that it was silly to fear death. After more than two thousand years his words still have not the slightest power to console us. Philip Larkin in *Aubade* was much more convincing about the fear of death:

> This is a special way of being afraid
> no trick dispels. Religion used to try,
> that vast moth-eaten musical brocade
> created to pretend we never die,
> and specious stuff that says *No rational being*
> *can fear a thing it will not feel,* not seeing
> that this is what we fear—no sight, no sound,
> no touch or taste or smell, nothing to think with,
> nothing to love or link with,
> the anaesthetic from which none come round.

Living without God isn't easy. But its very difficulty offers one other consolation—that there is a certain honor, or perhaps just a grim satisfaction, in facing up to our condition without despair and without wishful thinking—with good humor, but without God.

Sources

Index

Sources

Per Chapter

1. "Waiting for a Final Theory," *Time*, 155, 86 (April 10, 2000).

2. "Can Science Explain Everything? Anything?" *New York Review of Books*, 48, no. 9, 47–50 (May 31, 2001); reprinted in *Australian Review* (2001); reprinted in Portuguese in *Folha da S. Paolo* (2001); reprinted in French in *La Recherche* (2001); reprinted in *The Best American Science Writing 2002*, ed. M. Ridley and A. Lightman (HarperCollins, 2002); reprinted in *The Norton Reader* (W. W. Norton, New York, 2003); reprinted in *Explanations—Styles of Explanation in Science*, ed. John Cornwell (Oxford University Press, 2004): 23–38; reprinted in Hungarian in *Akadeemia* 176, No. 8, 1734–1749 (2005).

3. "Peace at Last in the Science Wars," in *The One Culture?* ed. J. A. Labinger and H. Collins (University of Chicago Press, Chicago, 2001): 238–240.

4. "The Future of Science, and the Universe," *New York Review of Books* XLVIII, no. 18, 58–63 (November 15, 2001).

5. "Dark Energy," *Times Higher Education Supplement*, December 21, 2001, VI.

6. "How Great Equations Survive," in *It Must Be Beautiful: Great Equations of Modern Science*, ed. G. Farmelo (Granta Books, London, 2002).

7. "On Missile Defense," *New York Review of Books* 49 no. 2, 41–47 (February 14, 2002); reprinted in *At Issue—Missile Defense* (Thomson—Gale, 2002); as course material by the U.S. Naval War College (2002); in *The Best American Science and Nature Writing 2003*, ed. Richard Dawkins (Houghton-Mifflin, Boston, 2003), 271–286.

8. "The Growing Nuclear Danger," *New York Review of Books* 49, no. 12, 18–20 (July 18, 2002).

9. "Is the Universe a Computer?" review of *A New Kind of Science*, by Stephen Wolfram, *New York Review of Books* 49, no. 16, 43–47 (October 24, 2002); reprinted in Spanish as "Es el universo un ordenador?" *Revista de Libros de la Fundación Caja Madrid*, no. 74, 23–30 (February 2003).

10. "Foreword to *A Century of Nature: Twenty-one Discoveries That Changed the Scientific World,*" ed. L. Garwin and T. Lincoln (University of Chicago Press, 2003), ix–x.

11. "Ambling toward Apocalypse," *Bulletin of the American Academy of Arts and Sciences* 56, no. 4, 35–44 (summer 2003); reprinted in *Federation of American Scientists Public Interest Report,* 56, no. 2 (summer 2003).

12. "What Price Glory?" *New York Review of Books* 50, no. 17, 55–60 (November 6, 2003); reply to R. H. Bloch, *New York Review of Books* 50, no. 20, 100–101 (December 18, 2003); reply to H. Lieber, *New York Review of Books* 51, no. 7, 57 (April 29, 2004).

13. "Four Golden Lessons," *Nature* 426, 389 (November 27, 2003).

14. "The Wrong Stuff," *New York Review of Books* 51, no. 6, 12–16 (April 8, 2004).

15. "A Turning Point?" contribution to symposium "The Election and America's Future," *New York Review of Books* 51, no. 17, 16–17 (November 4, 2004).

16. "About Oppenheimer," review of *Oppenheimer: Portrait of an Enigma,* by Jeremy Bernstein, *Physics Today,* January 2005, 51–52.

17. "Einstein's Search for Unification," in *Einstein—A Hundred Years of Relativity,* ed. Andrew Robinson (Palazzo Editions, Bath, England, 2005): 102–108.

18. "Einstein's Mistakes," *Physics Today* 58, no. 11, 31–35 (November 2005); reprinted in *E = Einstein—His Life, His Thought, and His Influence on Our Culture,* ed. D. Goldsmith and M. Bartusiak (Sterling, New York, 2006).

19. "Living in the Multiverse," opening talk, symposium "Expectations of a Final Theory," Trinity College, Cambridge, September 2, 2005; published in *Universe or Multiverse?* ed. B. Carr (Cambridge University Press, 2007), 29–42.

20. "Against the Boycott," invited opinion piece, *Times Higher Education Supplement,* May 26, 2006, 16.

21. "A Deadly Certitude," review of *The God Delusion,* by Richard Dawkins, in *Times Literary Supplement,* no. 5416, 5–6 (January 19, 2007).

22. "To the Postdocs," opinion piece, *Physics Today,* March 2007, 58; reprinted in Japanese, *Parity,* August 2007, 46–47.

23. "Science or Spacemen," *Sky and Telescope* 115, 96 (May 2008).

24. "Israel and the Liberals," *Tikkun,* May/June 2008, 86.

25. "Without God," Phi Beta Kappa Oration, Harvard University, June 3, 2008; *New York Review of Books* 55, 73 (September 25, 2008); excerpted in French in *Philosophie Magazine*, November 2008, 12; reprinted in Spanish in *Letras Libras*, March 2009, 12–17; reprinted in German in *Lettre International*, Spring 2009, 80–82.

Index